# ACE GENERAL CHEMISTRY I

*(THE EASY GUIDE TO ACE GENERAL CHEMISTRY I)*

*BY: DR. HOLDEN HEMSWORTH*

Copyright © 2015 by Holden Hemsworth

# DISCLAIMER

Chemistry, like any field of science, is continuously changing and new information continues to be discovered. The author and publisher have reviewed all information in this book with resources believed to be reliable and accurate and have made every effort to provide information that is up to date and correct at the time of publication. Despite our best efforts we cannot guarantee that the information contained herein is complete or fully accurate due to the possibility of the discovery of contradictory information in the future and any human error on part of the author, publisher, and any other party involved in the production of this work. The author, publisher, and all other parties involved in this work disclaim all responsibility from any errors contained within this work and from any results that arise from the use of this information. Readers are encouraged to check all information in this book with institutional guidelines, other sources, and up to date information.

*The information contained in this book is provided for general information purposes only and does not constitute medical, legal or other professional advice on any subject matter.* The author or publisher of this book does not accept any responsibility for any loss which may arise from reliance on information contained within this book or on any associated websites or blogs.

**Your reviews greatly help reach more students. If you find this book helpful, please leave a review on Amazon, nothing helps more than a few kind words.**

# WHY I CREATED THIS STUDY GUIDE

In this book, I try to breakdown the content covered in the typical first semester of a General Chemistry course in college for easy understanding and to point out the most important subject matter that students are likely to encounter. This book is meant to be a supplemental resource to lecture notes and textbooks to boost your learning and go hand in hand with your studying!

I am committed to providing my readers with books that contain concise and accurate information and I am committed to providing them tremendous value for their time and money.

Best regards,

Dr. Holden Hemsworth

# TABLE OF CONTENTS

# CHAPTER 1 – INTRODUCTION TO CHEMISTRY

## What is Chemistry?

*Chemistry is the branch of science concerned with the understanding of matter; the substances it is composed of and their properties, as well as the ways in which they interact and change to form new substances.*

## Matter

*Matter is anything that has mass and takes up space. Mass is the amount of matter an object contains; a way of quantifying matter.*

- Matter exists in three physical states

- Physical states of matter

    o Solid – matter with fixed shape and volume (rigid)

    o Liquid – matter with a fixed volume but indefinite shape

        ▪ Takes on the shape of the container it is in

    o Gas – matter without a fixed shape or volume

        ▪ Conforms to the volume and shape of its container

## Physical and Chemical Properties

- Physical property – characteristics that can be measured and observed without changing the chemical makeup of the substance

    o Examples: color, melting point, boiling point, density, etc.

- Physical change – a substance changes its physical appearance but does not change identity

    o Changes in state (e.g., liquid to gas, solid to liquid) are all physical changes

- Chemical property – any property that becomes evident during a chemical reaction

    o Examples: pH, corrosiveness, etc.

- Chemical change (aka chemical reactions) – a substance is transformed into a chemically different substance

## Mixtures

*Mixtures are combinations of two or more substances in which each substance keeps its chemical identity. Mixtures can be separated into two or more substances.*

- Heterogenous mixtures – mixture that is divided into different regions of appearance and properties
    - Results from components not being distributed uniformly
- Homogenous mixtures – mixture that is uniform throughout without any visible separations
    - Solutions are homogenous mixtures
        - Where a solid (the solute) is dissolved in a liquid (the solvent)

## Elements and Compounds

*Pure substances have definite and consistent composition and are composed of elements or compounds.*

- Element - substance that can't be broken down into other substances by chemical means
- Compound – substance formed from two or more chemical elements that are chemically bonded together
- Law of definite proportions
    - Pure compounds always contain exactly the same proportions of elements by mass

## Energy

*Energy is the capacity to do wok.*

- Kinetic energy – energy possessed by an object due to its motion
- Potential energy – energy stored in matter because of its position or location
    - Something suspended in the air has higher potential energy than something sitting on the ground
- Total Energy = potential energy + kinetic energy
- Lower energy states are more stable in nature

- Law of conservation of energy
  - Energy can't be created or destroyed…but it can be transformed
    - Example: potential energy can be converted to kinetic energy
  - Energy is always conserved

## Scientific Method

*The scientific method is a technique for investigation that is used to answer scientific questions.*

- Hypothesis - a proposed explanation made on observations or limited evidence that serves as the starting point for further investigation
- Theory – explanation of general principles of a phenomena that has been repeatedly tested and observed
- Fact – indisputable truth
- Steps in scientific approach
  - Observations
  - Hypothesis
  - Experiment
  - Development of a model or theory
  - Further experimentation

## Measurements

*Measured quantities consist of a number and a unit.*

- Units are standardized in the form of the International System called SI units
- Units have associated prefixes to make them easier to use and reports
  - Tera - $10^{12}$
  - Giga – $10^9$
  - Mega – $10^6$
  - Kilo – $10^3$
  - Deci – $10^{-1}$
  - Centi – $10^{-2}$
  - Milli – $10^{-3}$
  - Micro – $10^{-6}$
  - Nano – $10^{-9}$
  - Pico – $10^{-12}$

- Conversion factors – a mathematical multiplier used to convert a quantity expressed in one set of units into an equivalent quantity expressed in
  - Example: 1 yard = 3 feet (10 yard = 30 feet)

**Scientific Notation**

- Scientific notation is a way of handling very large or very small numbers
- The scientific notation for a number only contains significant figures
  - Examples
    - $525,000 = 5.25 \times 10^5$
    - $2,301,000,000 = 2.301 \times 10^9$
    - $0.000000000670 = 6.70 \times 10^{-10}$
- Consider the following: $0.000023 = 2.3 \times 10^{-5}$
  - The exponent on 10 is the number of places the decimal point must be shifted to give the number in its long form
    - Positive exponent, shift the decimal point to the right
    - Negative exponent, shift the decimal point to the left

**Significant Figures**

- All non-zero numbers are always significant
  - 1, 2, 3, 4, 5, 6, 7, 8, 9
- Zeroes in between non-zero numbers are always significant
  - 10001 – 5 sig figs
- A final zero or trailing zeroes in the **decimal portion** are significant
  - 0.00**500** – 3 sig figs
  - 1255.0 – 5 sig figs
- All zeroes to the left of a decimal point and that add value to a number are significant
  - 100.0 – 4 sig figs
  - 0.1 – 1 sig fig
    - Zero adds no value, it is there to avoid confusion and by convention

**Exact Number**

- Exact numbers are considered to have an infinite number of significant figures

  - They do not affect accuracy or precision of an expression they are in

  - You do not have to consider the significant figures in exact numbers when doing calculations

- Conversion factors are exact numbers

  - 1 yard = 3 feet (there are exactly 3 feet in a yard)

  - 1 foot = 12 inches (there are exactly 12 inches in a foot)

**Multiplication and Division Significant Figures**

- First perform all operations and arrive at an answer

- The answer should have the same number of significant figures as the number with the least amount of significant figures used in the calculations

**Addition and Subtraction Significant Figures**

- First perform all operations and arrive at an answer

- In addition and subtraction you only have to consider the significant figures in the decimal portion

  - The answer should contain no more decimal places than the number with the least amount of digits in the decimal portion

**Multiplication/Division Combined with Addition/Subtraction**

- Follow order of operations

- If the next operation to be performed is in the same group as the previous operation then don't round the calculation

  - For example when you perform division and then multiplication, you would not round the calculation

- If the next operation to be performed is in the other group from the previous operation then you would round the answer using the rules before moving on to the next operation

  - Example: You perform division and the next operation is subtraction

    - You would first round the result of the division using the significant figure rules for division before you perform subtraction

**Accuracy and Precision**

- Accuracy – how close a result is to the real value

- Precision – how close repeated measurements are in relation to one another

Accurate and Precise    Accurate but    Precise but
                    not Precise     not Accurate

*Accuracy vs. Precision:*

**Uncertainty**

- Uncertainty – error in a measurement

  - Expressed as a standard deviation

- When making a measurement involving an instrument, the measurement is made with one uncertain digit

20.1 mL

20.0 mL

19.9 mL

  - Example:

    - You record the measurement as 20.03

    - The 3 is an uncertain digit because it is estimated and can't be read off exactly from the instrument

**<u>Temperature</u>**

*Temperature is commonly quantified using the three units: kelvin, Celsius, and Fahrenheit.*

- Kelvin (K) – "absolute temperature scale"

  - Starts at absolute zero

  - Contains only positive values

- Celsius (°C) – "water based scale"

  - 0°C – freezing point of water

- 100°C – boiling point of water
- Most commonly used scale around the world
- Fahrenheit (°F) – "mercury based scale"
  - Commonly used in the US
- Converting Temperatures
  - Kelvin to Fahrenheit
    - (9/5)(K - 273) + 32
  - Kelvin to Celsius
    - K – 273
  - Celsius to kelvin
    - °C + 273
  - Celsius to Fahrenheit
    - (°C x 9/5) + 32
  - Fahrenheit to kelvin
    - (5/9)(°F – 32)  + 273
  - Fahrenheit to Celsius
    - (5/9)(°F - 32)

# CHAPTER 2 – COMPONENTS OF MATTER

## Components of Matter (Definitions)

- Element - substance that can't be broken down into other substances by chemical means

- Molecule - a combination of two or more atoms

- Compound – substance formed from two or more chemical elements that are chemically bonded together

- Mixture - two or more elements (or compounds) mingling without any chemical bonding

## Laws of Matter

- Law of Mass Conservation

  - Total masses of substances involved in a chemical reaction do not change

    - Number of substances and their properties can change

- Law of Definite Proportions:

  - Pure compounds contain exactly the same proportions of elements by mass

- Law of Multiple Proportions

  - If two elements react to form more than one compound, then the ratios of the masses of the second element which combine with a fixed mass of the first element will be in ratios of small whole numbers

## Postulates of Dalton's Atomic Theory

- All matter consists of extremely small particles called atoms

- All atoms of an element are identical

  - They are different from atoms of any other element

  - Including in mass and other properties

- Atoms of an element can't be converted into atoms of another element

- Compounds result when atoms of more than one element combine
  - A given compound has a specific ratio of atoms of different elements

## Periodic Table of Elements

*The periodic table is an arrangement of elements in rows and columns based on their atomic number, electron configurations, and chemical properties.*

- Period – horizontal row on the table

- Group (Family) – column on the table

| hydrogen 1 H 1.0079 | | | | | | | | | | | | | | | | | helium 2 He 4.0026 |
|---|---|---|---|---|---|---|---|---|---|---|---|---|---|---|---|---|---|
| lithium 3 Li 6.941 | beryllium 4 Be 9.0122 | | | | | | | | | | | boron 5 B 10.811 | carbon 6 C 12.011 | nitrogen 7 N 14.007 | oxygen 8 O 15.999 | fluorine 9 F 18.998 | neon 10 Ne 20.180 |
| sodium 11 Na 22.990 | magnesium 12 Mg 24.305 | | | | | | | | | | | aluminium 13 Al 26.982 | silicon 14 Si 28.086 | phosphorus 15 P 30.974 | sulfur 16 S 32.065 | chlorine 17 Cl 35.453 | argon 18 Ar 39.948 |
| potassium 19 K 39.098 | calcium 20 Ca 40.078 | scandium 21 Sc 44.956 | titanium 22 Ti 47.867 | vanadium 23 V 50.942 | chromium 24 Cr 51.996 | manganese 25 Mn 54.938 | iron 26 Fe 55.845 | cobalt 27 Co 58.933 | nickel 28 Ni 58.693 | copper 29 Cu 63.546 | zinc 30 Zn 65.39 | gallium 31 Ga 69.723 | germanium 32 Ge 72.61 | arsenic 33 As 74.922 | selenium 34 Se 78.96 | bromine 35 Br 79.904 | krypton 36 Kr 83.80 |
| rubidium 37 Rb 85.468 | strontium 38 Sr 87.62 | yttrium 39 Y 88.906 | zirconium 40 Zr 91.224 | niobium 41 Nb 92.906 | molybdenum 42 Mo 95.94 | technetium 43 Tc [98] | ruthenium 44 Ru 101.07 | rhodium 45 Rh 102.91 | palladium 46 Pd 106.42 | silver 47 Ag 107.87 | cadmium 48 Cd 112.41 | indium 49 In 114.82 | tin 50 Sn 118.71 | antimony 51 Sb 121.76 | tellurium 52 Te 127.60 | iodine 53 I 126.90 | xenon 54 Xe 131.29 |
| caesium 55 Cs 132.91 | barium 56 Ba 137.33 | 57-70 * | lutetium 71 Lu 174.97 | hafnium 72 Hf 178.49 | tantalum 73 Ta 180.95 | tungsten 74 W 183.84 | rhenium 75 Re 186.21 | osmium 76 Os 190.23 | iridium 77 Ir 192.22 | platinum 78 Pt 195.08 | gold 79 Au 196.97 | mercury 80 Hg 200.59 | thallium 81 Tl 204.38 | lead 82 Pb 208.98 | bismuth 83 Bi [209] | polonium 84 Po [210] | astatine 85 At [222] | radon 86 Rn |
| francium 87 Fr [223] | radium 88 Ra [226] | 89-102 ** | lawrencium 103 Lr [262] | rutherfordium 104 Rf [261] | dubnium 105 Db [262] | seaborgium 106 Sg [266] | bohrium 107 Bh [264] | hassium 108 Hs [269] | meitnerium 109 Mt [268] | ununnilium 110 Uun [271] | unununium 111 Uuu [272] | ununbium 112 Uub [277] | ununquadium 114 Uuq [289] | | | | |

| *Lanthanide series | lanthanum 57 La 138.91 | cerium 58 Ce 140.12 | praseodymium 59 Pr 140.91 | neodymium 60 Nd 144.24 | promethium 61 Pm [145] | samarium 62 Sm 150.36 | europium 63 Eu 151.96 | gadolinium 64 Gd 157.25 | terbium 65 Tb 158.93 | dysprosium 66 Dy 162.50 | holmium 67 Ho 164.93 | erbium 68 Er 167.26 | thulium 69 Tm 168.93 | ytterbium 70 Yb 173.04 |
|---|---|---|---|---|---|---|---|---|---|---|---|---|---|---|
| **Actinide series | actinium 89 Ac [227] | thorium 90 Th 232.04 | protactinium 91 Pa 231.04 | uranium 92 U 238.03 | neptunium 93 Np [237] | plutonium 94 Pu [244] | americium 95 Am [243] | curium 96 Cm [247] | berkelium 97 Bk [247] | californium 98 Cf [251] | einsteinium 99 Es [252] | fermium 100 Fm [257] | mendelevium 101 Md [258] | nobelium 102 No [259] |

- Elements on the periodic table can be classified as metals, nonmetals, and metalloids
  - Metal – substances that have luster, high heat conductivity, high electrical conductivity, and are solid at room temperature (exception: mercury)
  - Nonmetal – substance without any metal characteristics
  - Metalloid – substance that have both metal and nonmetal characteristics

## Atoms

*An atom is the smallest unit of matter.*

- Atoms interact to form molecules

- Atoms are composed of subatomic particles

  - Electrons – negatively charged particles

    - Carries a charge of -1.602 x $10^{-19}$ Coulombs (C)

      - Charge of atomic and sub-atomic particles are typically described as a multiple of this value

      - So, referred to as -1

    - Mass = 9.10938291 x $10^{-31}$ kg

  - Protons – positively charged particles

    - Carries a charge of +1.602 x $10^{-19}$ Coulombs (C)

      - Referred to as a +1 electron charge

    - Mass = 1.67262178 x $10^{-27}$

  - Neutrons – uncharged particles

    - Electrically neutral

    - Mass = 1.674927351 x $10^{-27}$

- Protons and neutrons are found in the nucleus

  - Nucleus is the central core of an atom

- Electrons orbit the nucleus in an "electron cloud"

- Elemental (atomic) symbol: shorthand representation of atoms of different elements

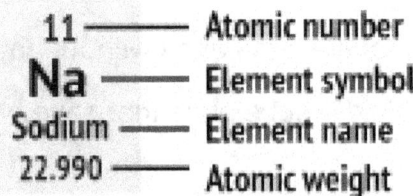

*Example of an Element on the Periodic Table:*

- Atomic number - number of protons in an atom of a particular element

  - For a neutral atom, number of electrons = number of protons

  - All atoms of an element have the same atomic number (same number of protons)

- Mass number = the number of protons + the number of neutrons

  - All atoms of an elements don't have the same number of neutrons

- Atomic weight (relative atomic mass) – average mass of atoms of an element

  - Calculated based on the relative abundance of isotopes in that particular element

  - Units: atomic mass units (amu)

- Isotopes – atoms of an element with the same number of protons but with a different number of neutrons

  - Same atomic mass but different mass number

## Types of Chemical Formulas

*Chemical formulas are a way of expressing information about the proportions of atoms that constitute a compound using: element symbols, numerical subscripts, and other symbols (e.g., parentheses, dashes).*

- Empirical formula – smallest whole number ratio of numbers of the atoms in a molecule

- Molecular formula – actual number of atoms in a molecule

- Structural formula – chemical formula showing how atoms are bonded together in a molecule

## Covalent and Ionic Bonds

### Covalent Bonds

- Two atoms share their valence electrons (electrons in the outer shell of an atom)

- Two Types

  - Non-polar covalent bond – electrons shared equally between atoms

    - Electronegativity of the two atoms is about the same

    - Typically electronegativity difference between the two atoms has to be less than 0.5 for non-polar bonds

- - Electronegativity – an atom's ability to attract and hold on to electrons, represented by a number
  - o Polar covalent bonds – electrons shared disproportionately between atoms
    - Electronegativity between the two atoms is different by a greater degree than 0.5 but less than 2.0

## Ionic Bonds

- Electrons are transferred, not shared between atoms
- An atom with high electronegativity will take an electron from an atom with low electronegativity
  - o Typically, difference in electronegativity is more than 2.0

## Ions

*Ions are charged atoms or molecules. Ions are formed when atoms or groups of atoms gain or lose valence electrons.*

- Monatomic ion – single atom with more or less electrons than the number of electrons in the atom's neutral state
- Polyatomic ions – group of atoms with excess or deficient number of electrons
- Anion – negatively charged ion
- Cation – positively charged ion
- Ionic compounds – association of a cation and an anion
  - o The cation is always named first

## Nomenclature

### Rules for Charges on Monoatomic Ions

- Elements in group 1 form monoatomic ions with charges equal to their group number
  - o Na is a group one element, forms $Na^+$, +1 charge
- Elements in group 2 form monoatomic ions with charges equal to their group number
  - o Mg is a group two element, forms $Mg^{2+}$, +2 charge

- Elements in group 17 form monoatomic ions with a -1 charge
  - Example: $Cl^-$, $F^-$, $I^-$

**Cations**

- Monatomic cations are formed from metallic elements
  - $Na^+$ - sodium ion
  - $Zn^{2+}$ - zinc ion
- Some elements can form more than one cation
  - The charge on the ion is indicated by a Roman numeral in parentheses followed by the name of the metal
    - $Fe^{2+}$ - iron (II) ion
    - $Fe^{3+}$ - iron (II) ion
  - Transition metals often form two or more different monoatomic cations

**Anions**

- Monoatomic anions are typically formed from nonmetals
  - Named by dropping the element name ending and adding –ide
    - $Cl^-$ - chloride ion
    - $F^-$ - fluoride ion
- Common polyatomic anions
  - OH – hydroxide ion
  - CN – cyanide ion
- Many polyatomic anions contain oxygen, they are called oxyanions
  - In elements that form two different oxyanions, the name of the one that contains more oxygen ends in -ate, the one with less ends in -ite:
    - $NO_2^-$ - nitrite ion
    - $NO_3^-$ - nitrate ion

- o Some compounds have multiple oxyanion forms
  - $ClO^-$ - hypochlorite ion, prefix "hypo" added to the oxyanion with the least number of oxygen, suffix "-ite"
  - $ClO_2^-$ - chlorite ion
  - $ClO_3^-$ - chlorate ion
  - $ClO_4^-$ - perchlorate ion, prefix "per" added to the oxyanion with the highest number of oxygen, suffix "-ate"
- o Many polyatomic anions with high (negative) charges can add one or more hydrogen cations (H+) to form anions with lower negative charge, their naming reflects whether the H+ addition involves one or more hydrogen ions
  - $HSO_4^-$ - hydrogen sulfate ion
  - $H_2PO_4^-$ - dihydrogen phosphate ion

**Acids**

*According to the Bronsted-Lowery definition, an acid is proton donating (donates H⁺).*

- Anions with names ending in -ide have associated acids that have the hydro-prefix and an -ic suffix:
  - o $Cl^-$ - chloride ion
    - $HCl$ – *hydro*chlor*ic* acid
- Acids of oxyanions
  - o If the oxyanion has an -ate ending, the corresponding acid is given an -ic ending
  - o If the oxyanion has an -ite ending, the corresponding acid has an -ous ending
  - o Prefixes used in the naming of the anion are kept in the name of the acid
    - $ClO^-$ - hypochlorite ion, $HClO$ - hypochlorous acid
    - $ClO_2^-$ - chlorite ion, $HClO_2$ - chlorous acid
    - $ClO_3^-$ - chlorate ion, $HClO_3$ chloric acid
    - $ClO_4^-$ - perchlorate ion, $HClO_4$ - perchloric acid

**Molecular Compounds**

- A pair of elements can form several different molecular compounds

  - Prefixes are used to identify the relative number of atoms in these compounds

    - $CO$ carbon **mono**xide

    - $CO_2$ carbon **di**oxide

- Prefixes

| | | | | | |
|---|---|---|---|---|---|
| o | 1: mono- | o | 5: penta- | o | 9: nona- |
| o | 2: di- | o | 6: hexa- | o | 10: deca- |
| o | 3: tri- | o | 7: hepta- | o | 11: undeca- |
| o | 4: tetra- | o | 8: octa- | o | 12: dodeca- |

**Hydrates**

- Hydrates are compounds that contains water molecules chemically bound to another compound or element

- Hydrates are first named from the anhydrous (dry) compound

  - It is then followed by the word "hydrate" and a prefix to indicate the number of water molecules

    - $CuSO_4 \cdot 5\ H_2O$ – copper (II) sulfate **pentahydrate**

*Monoatomic Cations and Anions*

| Common Cations | | | Transition Element Cations | | | Anions | | |
|---|---|---|---|---|---|---|---|---|
| Charge | Formula | Name | Charge | Formula | Name | Charge | Formula | Name |
| +1 | $H^+$ | Hydrogen | +2 | $Cd^{2+}$ | Cadmium | -1 | $H^-$ | Hydride |
| +1 | $Li^+$ | Lithium | +2 | $Cr^{2+}$ | Chromium(II) | -1 | $F^-$ | Fluoride |
| +1 | $Na^+$ | Sodium | +3 | $Cr^{3+}$ | Chromium(III) | -1 | $Cl^-$ | Chloride |
| +1 | $K^+$ | Potassium | +2 | $Mn^{2+}$ | Manganese (II) | -1 | $Br^-$ | Bromide |
| +1 | $Cs+$ | Cesium | +2 | $Fe^{2+}$ | Iron(II) | -1 | $I^-$ | Iodide |
| +2 | $Mg^{2+}$ | Magnesium | +3 | $Fe^{3+}$ | Iron(III) | -2 | $O_2^-$ | Oxide |
| +2 | $Ca^{2+}$ | Calcium | +2 | $Co^{2+}$ | Cobalt(II) | -2 | $S_2^-$ | Sulfide |
| +2 | $Sr^{2+}$ | Strontium | +3 | $Co^{3+}$ | Cobalt(III) | | | |
| +2 | $Ba^{2+}$ | Barium | +2 | $Ni^{2+}$ | Nickel(II) | | | |
| +3 | $Al^{3+}$ | Aluminum | +1 | $Cu^+$ | Copper(I) | | | |
| | | | +2 | $Cu^{2+}$ | Copper(II) | | | |
| | | | +2 | $Zn^{2+}$ | Zinc | | | |
| | | | +1 | $Hg_2^{2+}$ | Mercury(I) | | | |
| | | | +2 | $Hg^{2-}$ | Mercury(II) | | | |

*Polyatomic Ions:*

| Name | Formula | Name | Formula |
|---|---|---|---|
| Mercury(I) or mercurous | $Hg_2^{2+}$ | Nitrite | $NO_2^-$ |
| Ammonium | $NH_4^+$ | Nitrate | $NO_3^-$ |
| Cyanide | $CN^-$ | Hydroxide | $OH^-$ |
| Carbonate | $CO_3^{2-}$ | Peroxide | $O_2^{2-}$ |
| Hydrogen carbonate (or bicarbonate) | $HCO_3^-$ | Phosphate | $PO_4^{3-}$ |
| | | Monohydrogen phosphate | $HPO_4^{2-}$ |
| Acetate | $C_2H_3O_2^-$ | Dihydrogen phosphate | $H_2PO_4^-$ |
| Oxalate | $C_2O_4^{2-}$ | Sulfite | $SO_3^{2-}$ |
| Hypochlorite | $ClO^-$ | Sulfate | $SO_4^{2-}$ |
| Chlorite | $ClO_2^-$ | Hydrogen sulfite (or bisulfite) | $HSO_3^-$ |
| Chlorate | $ClO_3^-$ | | |
| Perchlorate | $ClO_4^-$ | Hydrogen sulfate (or bisulfate) | $HSO_4^-$ |
| Chromate | $CrO_4^{2-}$ | | |
| Dichromate | $Cr_2O_7^{2-}$ | Thiosulfate | $S_2O_3^{2-}$ |
| Permanganate | $MnO_4^-$ | | |

| Formula | Name |
|---|---|
| **Cations** | |
| $NH_4^+$ | **ammonium** |
| $H_3O^+$ | **hydronium** |
| **Anions** | |
| $CH_3COO^-$ (or $C_2H_3O_2^-$) | **acetate** |
| $CN^-$ | cyanide |
| $OH^-$ | **hydroxide** |
| $ClO^-$ | hypochlorite |
| $ClO_2^-$ | chlorite |
| $ClO_3^-$ | **chlorate** |
| $ClO_4^-$ | **perchlorate** |
| $NO_2^-$ | nitrite |
| $NO_3^-$ | **nitrate** |
| $MnO_4^-$ | **permanganate** |
| $CO_3^{2-}$ | **carbonate** |
| $HCO_3^-$ | **hydrogen carbonate** (or **bicarbonate**) |

| Formula | Name |
|---|---|
| **Cations** | |
| $NH_4^+$ | **ammonium** |
| $H_3O^+$ | **hydronium** |
| **Anions** | |
| $CrO_4^{2-}$ | chromate |
| $Cr_2O_7^{2-}$ | **dichromate** |
| $O_2^{2-}$ | peroxide |
| $PO_4^{3-}$ | **phosphate** |
| $HPO_4^{2-}$ | hydrogen phosphate |
| $H_2PO_4^-$ | dihydrogen phosphate |
| $SO_3^{2-}$ | sulfite |
| $SO_4^{2-}$ | **sulfate** |
| $HSO_4^-$ | hydrogen sulfate (or bisulfate) |

| Oxoanion | | Oxoacid | |
|---|---|---|---|
| $CO_3^{2-}$ | Carbonate ion | $H_2CO_3$ | Carbonic acid |
| $NO_2^-$ | Nitrite ion | $HNO_2$ | Nitrous acid |
| $NO_3^-$ | Nitrate ion | $HNO_3$ | Nitric acid |
| $PO_4^{3-}$ | Phosphate ion | $H_3PO_4$ | Phosphoric acid |
| $SO_3^{2-}$ | Sulfite ion | $H_2SO_3$ | Sulfurous acid |
| $SO_4^{2-}$ | Sulfate ion | $H_2SO_4$ | Sulfuric acid |
| $ClO^-$ | Hypochlorite ion | $HClO$ | Hypochlorous acid |
| $ClO_2^-$ | Chlorite ion | $HClO_2$ | Chlorous acid |
| $ClO_3^-$ | Chlorate ion | $HClO_3$ | Chloric acid |
| $ClO_4^-$ | Perchlorate ion | $HClO_4$ | Perchloric acid |

## Chemical Equations

- Chemical reactions are expressed through chemical equations
- An arrow ("→") in a chemical equation means "yields"
  - $2 H_2(g) + O_2(g) \rightarrow 2 H_2O(l)$
    - Hydrogen + oxygen yields water
  - $H_2$ and $O_2$ are reactants
    - Substances that undergo change during a reaction
  - $H_2O$ is the product
    - Substances formed from chemical reactions
- Common phase notation
  - g = gas
  - l = liquid
  - s = solid

## Balancing Chemical Equations

- Balanced chemical equations adhere to the Law of Conservation of Matter
  - A balanced equation has to have equal numbers of each type of atom on both sides of the arrow
- Balancing is done by changing the coefficients
  - The coefficient times the subscript gives the total number of atoms
  - If there are no coefficients in front, coefficient is equal to one
  - If an atom doesn't have a subscript, subscript is equal to one
- Subscripts are **never** changed

# CHAPTER 3 – STOICHIOMETRY OF FORMULAS AND EQUATIONS

## Mass and Moles

- In the metric system, the standard unit of mass is the gram (or kilogram)

- All elements have a unique mass (atomic weight)

    - Expressed as either atomic mass units (amu) or grams

    - Same weight of two different elements represents a different number of atoms

- Consider the reaction: $H_2 + F_2 \rightarrow 2\ HF$

    - Does not mean that 1 gram of hydrogen will react with 1 gram of fluorine to form 2 grams of hydrogen fluoride

    - In reality 2.016 g of Hydrogen will react with 38.000 g of Fluorine to form 40.016 g Hydrogen Fluoride

        - 2.016 g of Hydrogen contain the same number of $H_2$ molecules as 38.000 g of Fluorine ($F2$)

        - 40.016 grams of HF will contain twice as many molecules

- Number of molecules, even in low masses, are extremely large numbers

    - So for convenience, amounts in chemistry are expressed in moles

- Mole - quantity of a substance that contains the same number of atoms, molecules or formula units as exactly 12 g of Carbon-12

    - 1 mole (mol) = $6.0221 \times 10^{23}$

- Atomic mass – mass of one molecule

    - Expressed in atomic mass units (amu)

- Molar mass – mass of one mole of entities (atoms, molecules, formula units) of a substance

    - Expressed in g/mole

- Molar mass and atomic mass are numerically similar
    - Example: one molecule of Carbon has an atomic mass of 12.0107 amu and a molar mass of 12.0107 g/mol
        - In 12.0107 g of Carbon there are $6.0221 \times 10^{23}$ molecules

## Mass Percentage (Percent Composition)

*Mass percentage is a way of expressing the concentration of an element in a compound or a compound of a mixture. Steps for solving percent composition (aka mass percentage) questions:*

Example Question: Find the mass percentages of C, O, and H in glucose ($C_6H_{12}O_6$)

- First, look up the atomic masses of the elements that are in the compound on a periodic table
    - C – 12.01 g
    - H – 1.01 g
    - O – 16.00 g
- Second, determine how many grams of each element are in one mole of glucose (or whatever compound a question may be asking you for)
    - C – (6 moles of C x 12.01 g) = 72.06 g
    - H – (12 moles of H x 1.01 g) = 12.12 g
    - O – (6 moles of O x 16.00 g) = 96.00 g
- Third, determine the total mass in one mole of the compound by adding up the masses of the elements from step 2
    - Mass of one mole of glucose = 180.18g (72.06 g + 12.12 g + 96.00 g)
- Finally, find the mass percentages of the elements by dividing the weight of each element in one mole of the compound by the molar mass of that compound
    - C – (72.06 g / 180.18) x 100% = 39.99%
    - H – (12.12 g / 180.18) x 100% = 6.73%
    - O – (96.00 g/ 180.18) x 100% = 53.28%
- To check your work you can add up the percentages to see if they add up to 100%
    - 39.99% + 6.73% + 53.28% = 100%

**Formula**

$$Mass\ \%\ of\ Element\ Z$$

$$= \frac{moles\ of\ Z\ in\ formula\ *\ molar\ mass\ of\ Z\ \left(\frac{g}{mol}\right)}{mass\ (g)\ of\ 1\ mole\ of\ compound}$$

## Determining Empirical Formula

*Empirical formulas are the smallest whole number ratio of numbers of the atoms in a molecule. The molecular formula of a compound is the formula of the compound as it exists, and may be a multiple of the empirical formula.*

### Determining Empirical Formula from Masses

Example Question: A compound contains 36.42 g of Carbon, 6.12 g of Hydrogen, and 47.89 g of Oxygen, what is its empirical formula?

- First, determine the moles of each element

    o C – (36.42 / 12.01) = 3.03

    o H – (6.12 / 1.01) = 6.06

    o O – (47.89 / 16.00) = 2.99

- Second, determine the lowest whole-number ratios; divide the moles of each element by the lowest mole amount

    o C – (3.03 / 2.99) = 1.01 → 1

    o H – (6.06 / 2.99) = 2.03 → 2

    o O – (2.99 / 2.99) = 1.00 → 1

      ▪ This step usually results in ratios that are very close to a whole number

      ▪ However, in some question you may get ratios of 1.5, or 2.5, or 3.5, etc. in this case you would multiply all the ratios by 2 to get whole number ratios

      ▪ In some question you may get ratios of 1.33, or 2.33, or 3.33, etc. in this case you would multiply all the ratios by 3 to get whole number ratios

      ▪ In general terms, if the ratios are not very close to a whole number you have to multiply them by a number that would result in approximately whole numbers

- Write the empirical formula from the results

  - $CH_2O$

## Determining Empirical Formula from Elemental Analysis (% Composition)

Example Question: A compound is found to contain 56% Carbon, 7% Hydrogen, and 37% Oxygen. What is the empirical formula for this compound? The molecular weight for this compound is 86.14 g/mol. What is the molecular formula?

- First, assume exactly 100 g of the compound is present

  - This allows you to exchange percentages with grams

    - C – 56 g

    - H – 7 g

    - O – 37 g

- Second, convert masses to moles

  - C – (56 / 12.01) = 4.66

  - H – (7 / 1.01) = 6.93

  - O – (37 / 16.00) = 2.31

- Third, determine the lowest whole-number ratios; divide the moles of each element by the lowest mole amount

  - C – (4.66 / 2.31) = 2.02 → 2

  - H – (6.93 / 2.31) = 3.00 → 3

  - O – (2.31 / 2.31) = 1.00 → 1

- Write the empirical formula from the results

  - $C_2H_3O$

- To determine the molecular formula from the empirical formula follow these steps:

  o Calculate the weight from the empirical formula (multiply atoms of each element with the elements molar mass and add them up)

    - 2 Carbon atoms x 12.01 g = 24.02 g

    - 3 Hydrogen atoms x 1.01 g = 3.03 g

    - 1 Oxygen atom x 16.00 g = 16.00 g

    - Total : 24.02 g + 3.03 g + 16.00 g = 43.05 g

  o Divide the molecular weight by the weight determined from the empirical formula to find the scaling factor

    - 86.14 / 43.05 = 2.00

    - Scaling factor is 2

  o Using the scaling factor determine the molecular formula

    - $C_4H_6O_2$

## Stoichiometry

*Stoichiometry involves using relationships between elements, compounds, chemical formulas, and chemical reactions to acquire quantitative data. There are four major categories of stoichiometry problems that you are likely to encounter. They are listed below with strategies on how to solve them.*

- To convert from the mass of a substance to moles of that substance you divide by the molar mas

- To convert from moles of a substance to the mass of a substance you multiply by the molar mass

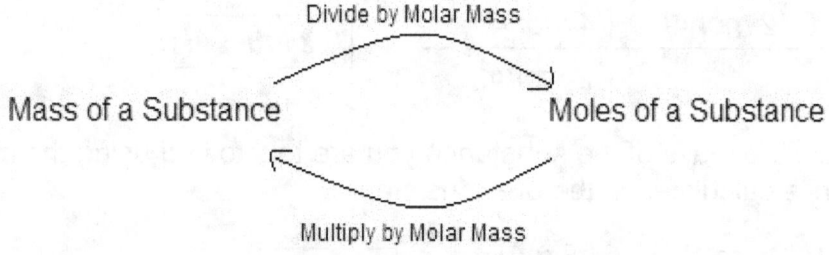

Divide by Molar Mass

Mass of a Substance          Moles of a Substance

Multiply by Molar Mass

*Interconversion:*

- This interconversion is very important in chemical calculations

**Stoichiometric Mole–Mole Problems**

Example Question: How many moles of HCl are needed to react with 0.82 moles of Al?

- Write out a chemical equation from the information given in the question
  - $Al + HCl \rightarrow AlCl_3 + H_2$

- Balance the chemical equation
  - $2\,Al + 6\,HCl \rightarrow 2\,AlCl_3 + 3\,H_2$

- Calculate the moles of the substance you are told to find using mole ratios
  - $$\frac{0.82\ mol\ Al \mid 6\ mol\ HCl}{2\ mol\ Al} = \boxed{2.46\ mol\ HCl}$$

**Stoichiometric Mass–Mass Problems**

Example Question: How many grams of Al can be created from decomposing 12.60 g of $Al_2O_3$?

- Write out a chemical equation from the information given in the question
  - $Al_2O_3 \rightarrow Al + O_2$

- Balance the chemical equation
  - $2\,Al_2O_3 \rightarrow 4\,Al + 3\,O_2$

- Convert the mass of the given substance into moles of the given (mol/g)
  - $$\frac{12.6\ g\ Al_2O_3 \mid 1\ mol\ Al_2O_3}{101.96\ g\ Al_2O_3} = \boxed{0.12\ mol\ Al_2O_3}$$

- Calculate the moles of the substance you are trying to find the mass for using mole ratios
  - $$\frac{0.12\ mol\ Al_2O_3 \mid 4\ mol\ Al}{2\ mol\ Al_2O_3} = \boxed{0.24\ mol\ Al}$$

- Calculate the mass of the substance you are told to find using the moles of the substance calculated in the previous step
  - $$\frac{0.24\ mol\ Al \mid 26.98\ g\ Al}{1\ mol\ Al} = \boxed{6.48\ g\ Al}$$

**Stoichiometric Mass-Volume Problems**

Example Question: How many liters of $H_2$ are created from the reaction of 40.00 g of K in water?

*Important*: 1 mole of any ideal gas will occupy 22.41 L

- Write out a chemical equation from the information given in the question

    o $K + H_2O \rightarrow KOH + H_2$

- Balance the chemical equation

    o $2 K + 2 H_2O \rightarrow 2 KOH + H_2$

- Convert the mass of the given substance into moles of the given (mol/g)

    o $$\frac{40.0 \, g \, K}{} \Big| \frac{1 \, mol \, K}{39.10 \, g \, K} = \boxed{1.02 \, mol \, K}$$

- Calculate the moles of the substance you are trying to find the volume for using mole ratios

    o $$\frac{1.02 \, mol \, K}{} \Big| \frac{1 \, mol \, H_2}{2 \, mol \, K} = \boxed{0.51 \, mol \, H_2}$$

- Calculate the volume of the substance you are told to find using the moles of the substance calculated in the previous step

    o $$\frac{0.51 \, mol \, H_2}{} \Big| \frac{22.41 \, L \, H_2}{1 \, mol \, H_2} = \boxed{11.43 \, L \, H_2}$$

**Stoichiometric Volume-Volume Problems**

Example Question: How many liters of $SO_2$ will be produced from 28.3 L $O_2$?

*Important*: 1 mole of any ideal gas will occupy 22.41 L

- Write out a chemical equation from the information given in the question

    o $S_2 + O_2 \rightarrow SO_2$

- Balance the chemical equation

    o $S_2 + 2 O_2 \rightarrow 2 SO_2$

- Convert the volume of the given substance into moles of the given (mol/g)

$$\frac{28.3\ L\ O_2}{} \bigg| \frac{1\ mol\ O_2}{22.41\ L\ O_2} = \boxed{1.26\ mol\ O_2}$$

- Calculate the moles of the substance you are trying to find the volume for using mole ratios

$$\frac{1.26\ mol\ O_2}{} \bigg| \frac{2\ mol\ SO_2}{2\ mol\ O_2} = \boxed{1.26\ mol\ SO_2}$$

- Calculate the volume of the substance you are told to find using the moles of the substance calculated in the previous step

$$\frac{1.26\ mol\ SO_2}{} \bigg| \frac{22.41\ L\ SO_2}{1\ mol\ SO_2} = \boxed{28.24\ L\ SO_2}$$

## Limiting Reagent

- Typically one of the reactants in a chemical reaction is present in smaller stoichiometric amounts than the other reactant(s)

  o This limits the amount of product(s) that can form

- Limiting reagent problems are easy to identify

  o The problem will give the amounts of more than one of the starting materials

  o You will have to determine the limiting reagent from calculations in order to solve the problem

- There are two typical methods of determining the limiting reagent

  o Need vs. Have Method

  o Product Method

### Need vs. Have Method

- Pick one of the reactants (Reac1), calculate how much of the other reactant (Reac2) you will need to completely react with Reac1

  o Utilize stoichiometric ratios

    ▪ This requires a balanced equation!

    ▪ Make sure the equation is balanced before you begin working on the problem

- Compare the amount needed for the Reac2 with the actual amount listed in the question

  - If the amount you need is MORE than what you actually have available according to the problem then Reac2 is the limiting reagent

  - If the amount you need is LESS than what you actually have available according to the problem then Reac1 is the limiting reagent

- You finish the rest of the problem based on the reactant you found to be limiting

## "Product" Method

- Pick one of the products of the reaction

  - Choose one of the reactants (Reac1) and calculate how much of the product you can make using the amount you have available according to the question

  - Make the same calculation using the other reactant (Reac2)

- The reactant that results in the smallest amount of product is the limiting reagent

  - The amount is also the maximum amount of product that can be made

    - This amount is called the "theoretical yield"

## Reagent in Excess

- Some questions may ask you to determine how much of the excess reagent will be leftover once the reaction ends

- To calculate this value

  - Calculate how much of the excess reagent is needed to completely react with the limiting reagent

  - Take the difference (subtract) of the amount of excess reagent you had at the beginning and the amount needed to completely react with the limiting reagent

## Percent Yield

- It is very common that the amount of product made experimentally is lower than the amount expected by the theoretical yield

  - This occurs because of mechanical errors, incomplete reactions, "side" reactions, etc.

- To determine the percent yield (sometimes also called the efficiency of a reaction) use the following formula:

  - $\%\ Yield = \dfrac{actual\ yield}{theoretical\ yield}\ x\ 100\%$

## Solutions

- Solutions are composed of a solute and a solvent

  - Solutions are homogeneous mixtures

  - Solutes are minor components in a solution (present in smaller amounts)

  - Solvents are substances in which a solute is dissolved

- Solution composition is expressed by:

  - Mass percent

    - $mass\ \% = \dfrac{mass\ solute}{mass\ of\ solution}\ x\ 100\%$

    - $mass\ \% = \dfrac{mass\ solute}{mass\ solvent\ +\ mass\ solute}\ x\ 100\%$

  - Mole fraction

    - $X_{solute} = \dfrac{mass\ solute}{mass\ solvent\ +\ mass\ solute}$

    - $X_{solute} = n_{solute}\ /\ n_{total}$

  - Molarity

    - $M = \dfrac{moles\ solute}{Liters\ of\ solution} = \dfrac{mol}{L}$

  - Molality

    - $m = \dfrac{moles\ solute}{kg\ solvent} = \dfrac{mol}{kg}$

**Concentrated vs. Dilute Solutions**

- Concentrated solution – a solution that contains a high amount of solute

    o More solute particles per unit volume

- Dilute solution – a solution that contains a low amount of solute

    o Fewer solute particles per unit volume

- Saturated Solution - a solution that contains the maximum amount of solute that can be dissolved by the solvent

**Dilutions**

- Extra solvent is added to a solution to dilute it

    o The amount of solute in the solution remains the same

- Use the following formula to solve dilution problems:

    o $M_1V_1 = M_2V_2$

        ▪ $M_1$ – is the concentration of stock solution

        ▪ $V_1$ – is the volume of stock solution

        ▪ $M_2$ – is the concentration of the final solution

        ▪ $V_2$ – is the volume of the final solution

# CHAPTER 4 – CHEMICAL REACTIONS

## Water as a Solvent

- Most reactions that occur in organisms and the environment take place in water

- Water has 2 hydrogen atoms covalently bonded to 1 oxygen atom

- The electronic structure of water is tetrahedral

  - 2 covalent bonds with H atoms and 2 sets of unpaired electrons

  - 104.5° bond angle – because the two lone pairs try to separate as far as possible

- Water is a **polar** molecule

  - Polar atoms have dipoles because of unequal sharing of electrons

    - The shared pair of electrons are attracted more strongly to the Oxygen forming a partially negative charged "pole" near the Oxygen and partially positively charged poles near the Hydrogens

*Structure of Water with Dipoles:*

## Solubility

- Soluble substances (solutes) in water

  - Interaction between an ion and water is strong

  - Polar compounds are water soluble

    - Polar water molecules have an affinity for oppositely charged regions of other polar molecules

- Insoluble substances (solutes) in water
  - Non-polar compounds are not water-soluble
  - If the interaction between ion and water is weak
    - Insoluble substances have a force of attraction so strong in their solid substance form, that it cannot be overcome by the interaction of the ions with the polarized water molecules

## Ions in Aqueous Solution

- In ionic solids, cations & anions are held together by electrostatic attraction
  - The electrostatic attraction is replaced by water molecules in an aqueous solution
- Many ionic compounds dissociate into independent ions when dissolved in water
  - Compounds that freely dissociate into independent ions in aqueous solution are called electrolytes
    - Solutions with electrolytes are good conductors of electricity
    - Not all electrolytes are ionic compounds
- Some molecular compounds dissolve but do not dissociate into ions
  - Contain polar bonds which interact with the polar bonds in water
  - Do not conduct an electric current
    - Produce a nonconducting solution
  - Referred to as nonelectrolytes

## Strong and Weak Electrolytes

- Strong electrolyte - an electrolyte that completely disassociates in solution
  - Exist in solution almost entirely as ions
- Weak electrolyte - an electrolyte that does not completely disassociate in solution
  - Exists in solution as both ions and molecules of the electrolyte

## Molecular and Ionic Equations

- Molecular equation express reactants and products as if they were molecules, despite actually existing in solution as ions

  - Example of a molecular equation

    - $CaCl_2$ (aq) + 2 $AgNO_3$ (aq) $\rightleftharpoons$ $Ca(NO_3)_2$ (aq) + 2 $AgCl$ (s)

    - (aq) is used to indicate that the substance is actually disassociated in solution

    - (s) is used to indicate that the substance is a precipitate

- Ionic equation express strong electrolytes as separate independent ions

  - Example of an ionic equation

    - $Ca^{2+}$ (aq) + 2 $Cl^-$ (aq) + 2 $Ag^+$ (aq) + 2 $NO_3^-$ (aq) $\rightleftharpoons$ $Ca^{2+}$ (aq) +2 $(NO_3)^-$ (aq) + 2 $AgCl$ (s)

    - Note that the solid is written in its full formula

- Net ionic equation are equations without any spectator ions

  - Spectator ion - an ion in an ionic equation that does not take part in the reaction

  - Easy way of identifying spectator ions is by writing the ionic equation and then crossing off any ions that appear on both sides of the equation (those are the spectator ions)

    - Example:

    - ~~$Ca^{2+}$ (aq)~~ + 2 $Cl^-$ (aq) + 2 $Ag^+$ (aq) + ~~2 $NO_3^-$ (aq)~~ $\rightleftharpoons$ ~~$Ca^{2+}$ (aq)~~ ~~+2 $(NO_3)^-$ (aq)~~ + 2 $AgCl$ (s)

  - Example of a net ionic equation

    - 2 $Cl^-$ (aq) + 2 $Ag^+$ (aq) $\rightleftharpoons$ 2 $AgCl$ (s)

## Three Major Classes of Chemical Reactions

*The three major classes of chemical reactions are: precipitation reactions, acid-base reactions, and oxidation-reduction reactions.*

### Precipitation Reactions

- Precipitation reaction occurs in aqueous solution because one of the product in a precipitation reaction is insoluble

    o Precipitate - an insoluble solid compound formed during a chemical reaction in solution

- There are generalized solubility rules that are used to predict whether a precipitate will form

    o Soluble - all compounds containing the ammonium ion ($NH_4^+$) or alkali metal (Group IA on the periodic table) cations

    o Soluble - all nitrates and acetates (ethanoates)

    o Soluble - all chlorides, bromides and iodides

        ▪ Except those with Ag, Pb, Hg

    o Soluble - all sulfates

        ▪ Except those with Ag, Pb, Hg(I), Ba, Sr, Ca

    o Insoluble - all carbonates, sulfites, and phosphates

        ▪ Except those with ammonium ($NH_4^+$), and alkali metal (Group IA) cations

    o Insoluble - all Hydroxides

        ▪ Except those with $NH_4^+$, alkali metal (Group IA) cations

    o Insoluble - all sulfides

        ▪ Except those with $NH_4^+$, alkali metal (Group Ia) cations, and alkali earth metal (Group II) cations

    o Insoluble - all oxides

        ▪ Except those with Calcium, Barium, and alkali metal (group IA) cations

*Solubility Rules Tables:*

## Solubility Rules for Ionic Compounds

| Rule | Applies to | Statement | Exceptions |
|------|-----------|-----------|------------|
| 1 | $Li^+$, $Na^+$, $K^+$, $NH_4^+$ | Group IA and ammonium compounds are soluble. | — |
| 2 | $C_2H_3O_2^-$, $NO_3^-$, $ClO_4^-$ | Acetates and nitrates are soluble. | — |
| 3 | $Cl^-$, $Br^-$, $I^-$ | Most chlorides, bromides, and iodides are soluble. | $AgCl$, $Hg_2Cl_2$, $PbCl_2$, $AgBr$, $HgBr_2$, $Hg_2B$ $PbBr_2$, $AgI$, $HgI_2$, $Hg_2I_2$, $PbI_2$ |
| 4 | $SO_4^{2-}$ | Most sulfates are soluble. | $CaSO_4$, $SrSO_4$, $BaSO_4$, $Ag_2SO_4$, $Hg_2SO$ $PbSO_4$ |
| 5 | $CO_3^{2-}$ | Most carbonates are insoluble. | Group IA carbonates, $(NH_4)_2CO_3$ |
| 6 | $PO_4^{3-}$ | Most phosphates are insoluble. | Group IA phosphates, $(NH_4)_3PO_4$ |
| 7 | $S^{2-}$ | Most sulfides are insoluble. | Group IA sulfides, $(NH_4)_2S$ |
| 8 | $OH^-$ | Most hydroxides are insoluble. | Group IA hydroxides, $Ca(OH)_2$, $Sr(OH)$ $Ba(OH)_2$ |

| Negative Ions (Anions) | + | Positive Ions (Cations) | = | Solubility of Compounds In Water | Example |
|---|---|---|---|---|---|
| Any Anion | + | Alkali Ions ($Li^+$, $Na^+$, $K^+$, $Rb^+$, $Cs^+$, $Fr^+$) | = | Soluble | NaF, $KNO_3$ |
| Any Anion | + | Hydrogen Ion [$H^+$(aq)] | = | Soluble | HCl |
| Any Anion | + | Ammonium Ion ($NH_4^+$) | = | Soluble | $NH_4Cl$ |
| Nitrate ($NO_3$) | + | Any Cation | = | Soluble | $Ca(NO_3)_2$ |
| Acetate ($CH_3COO^-$) | + | Any Cation | = | Soluble | $CH_3COONa$ |
| Halides ($Cl^-$, $Br^-$, $I^-$) | + | $Ag^+$, $Pb^{2+}$, $Hg^{2+}$, $Cu^+$, $Tl^+$ | = | Insoluble | AgCl, $PbBr_2$ |
| | + | Any Other Cation | = | Soluble | KBr, $CaI_2$ |
| Sulfate ($SO_4^{2-}$) | + | $Ca^{2+}$, $Sr^{2+}$, $Ba^{2+}$, $Ag^+$, $Pb^{2+}$, $Ra^{2+}$ | = | Insoluble | $BaSO_4$ |
| | + | Any Other Cation | = | Soluble | $CuSO_4$ |
| Sulfide ($S^{2-}$) | + | Alkali Ions – $Li^+$, $Na^+$, $K^+$, $Rb^+$, $Cs^+$, $Fr^+$ Alkali Earth Metals – $Be^{2+}$, $Mg^{2+}$, $Ca^{2+}$, $Sr^{2+}$, $Ba^{2+}$, $Ra^{2+}$, $H^+$(aq), $NH_4^+$ | = | Soluble | $H_2S$, MgS, $(NH_4)_2S$ |
| | + | Any Other Cation | = | Insoluble | ZnS |
| Hydroxide ($OH^-$) | + | Alkali Ions - $Li^+$, $Na^+$, $K^+$, $Rb^+$, $Cs^+$, $Fr^+$, $H^+$(aq), $NH_4^+$, $Sr^{2+}$, $Ba^{2+}$, $Ra^{2+}$, $Tl^+$ | = | Soluble | $Sr(OH)_2$ |
| | + | Any Other Cation | = | Insoluble | AgOH |
| Phosphate ($PO_4^{3-}$) Carbonate ($CO_3^{2-}$) | + | Alkali Ions - $Li^+$, $Na^+$, $K^+$, $Rb^+$, $Cs^+$, $Fr^+$, $H^+$(aq), $NH_4^+$ | = | Soluble | $(NH_4)_3PO_4$ |
| | + | Any Other Cation | = | Insoluble | $MgCO_3$ |

- Precipitation reactions take the form of an "exchange reaction"
  - Exchange reactions are reaction between compounds that appear to involve an exchange of cations and anions

## Acid-Base Reactions

- The Arrhenius definition of acids and bases
  - Acids produce Hydrogen ions ($H^+$) when dissolved in water
  - Bases produce Hydroxide ions ($OH^-$) when dissolved in water
- Bronsted-Lowery definition of acids and bases

- o Acids are proton donating

- o Bases are proton accepting

- Strong acid – an acid that ionizes completely in water

  - o Strong acids are $HI$, $HBr$, $HClO_4$, $HCl$, $HClO_3$, $H_2SO_4$, and $HNO_3$

- Weak acid – an acid that only partially ionizes in water

- Strong base – a base that is present almost entirely as ions (one of the ions is $OH^-$)

  - o Strong bases are $NaOH$, $KOH$, $LiOH$, $RbOH$, $CsOH$, $Ca(OH)_2$, $Ba(OH)_2$, and $Sr(OH)_2$

- Weak base – a base that only partially ionizes in water

- Strong acids and bases are represented as separate ions in an ionic equation

- Weak acids and bases are represented as undissociated "molecules" in ionic equations

  - o Example of Acetic Acid (a weak acid)

  - o Molecular Equation: $CH_3COOH_{(aq)} + NaOH_{(aq)} \rightarrow CH_3COONa_{(aq)} + H_2O_{(l)}$

  - o Ionic: $CH_3COOH_{(aq)} + Na^+_{(aq)} + OH^-_{(aq)} \rightarrow CH_3COO^-_{(aq)} + Na^+_{(aq)} + H_2O_{(l)}$

  - o Net: $CH_3COOH_{(aq)} + OH^-_{(aq)} \rightarrow CH_3COO^-_{(aq)} + H_2O_{(l)}$

**Neutralization (Acid-Base) Reactions**

- Acids and bases neutralize one another

- Neutralization reaction - reaction between an acid and a base that results in an ionic compound and water

- Salt - ionic compound that is the product of a neutralization reaction

- Net ionic equation for each neutralization reaction involves a transfer of a proton

  - o Consider the example of HCl (a strong acid) reacting with LiOH (a strong base)

    - ▪ Net equation for the reaction is: $H^+_{(aq)} + OH^-_{(aq)} \rightarrow H_2O_{(l)}$

  - o Consider the example of HCN (a weak acid) reacting with KOH (a weak base)

    - ▪ $HCN_{(aq)} + OH^-_{(aq)} \rightarrow CN^-_{(aq)} + H_2O_{(l)}$

- The proton is transferred from HCN to OH⁻

## Acid-Base Reactions with Gas Formation (aka Displacement Reactions)

- Acid-base reactions with gas formation sometimes involve unstable chemical species (e.g., $H_2CO_3$ and $H_2SO_3$

- Unstable species are enclosed in parentheses

    o Example: $[H_2CO_{3 (aq)}] \rightarrow H_2O_{(l)} + CO_{2 (g)}$

## Oxidation-Reduction Reactions (aka Redox Reactions)

- Oxidation-reduction reactions (aka redox reactions) – reactions that involve a partial or complete transfer of electrons from one reactant to another

    o Oxidation = loss of electrons

    o Reduction = gain of electrons

        ▪ Trick for remembering which is which - OIL RIG

        ▪ **OIL - O**xidation **I**s **L**osing electrons

        ▪ **RIG - R**eduction **I**s **G**aining electrons

    o Oxidation and reduction always occur simultaneously

- Oxidation number (aka oxidation state) – actual charge an atom in a molecule would have if all the electrons it was sharing were transferred completely, not shared

- Oxidizing agent – species that oxidizes another species

    o It is itself reduced

- Reducing agent – species that reduces another species

    o It is itself oxidized

*Reducing and Oxidizing Agents:*

| Applies to | Rule |
|---|---|
| Elements | The oxidation number of an atom in an element is zero. |
| Monatomic ions | The oxidation number of an atom in a monatomic ion equals the charge of the ion. |
| Oxygen | $-2$ in oxides, e.g. $ZnO$, $CO_2$, $H_2O$<br>$-1$ in peroxides, e.g. $H_2O_2$<br>$-(1/2)$ in superoxides, e.g. $KO_2$<br>$-(1/3)$ in inorganic ozonides, e.g. $RbO_3$<br>$0$ in $O_2$<br>$+(1/2)$ in dioxygenyl, e.g. dioxygenyl hexafluoroarsenate $O_2^+ [AsF_6]^-$<br>$+1$ in $O_2F_2$<br>$+2$ in $OF_2$ |
| Hydrogen | The oxidation number of hydrogen is $+1$ in most of its compounds |
| Halogens | Fluorine is $-1$ in all its compounds. The other halogens are $-1$ unless the other element is another halogen or oxygen |
| Compounds and ions | The sum of the oxidation numbers of the atoms in a neutral compound is zero. The sum in a polyatomic ion equals the charge on the ion |
| Carbon in Organic Compounds | The oxidation number of Carbon in Organic Compounds has an average value of 0<br>The O.N. of other elements in carbon compounds ($H^{+1}$, $O^{-2}$, $N^{+3}$, $Cl^{-1}$, etc.) still apply |

- Common Redox Reactions

  - Combination - reaction in which two substance combine to form a third substance

  - Decomposition – reaction in which a single compound produces two or more substances

  - Displacement (aka single-replacement reaction) – reaction in which an element reacts with a compound and displaces an element from it

  - Combustion – reaction in which one of the reactants is oxygen

    - Usually results in the rapid release of heat

- Redox reactions can be written in terms of two half-reactions

  - One involves the loss of electrons (oxidation)

  - The other involves the gain of electrons (reduction)

  - Example: $Fe^{2+} + Ce^{4+} \rightarrow Fe^{3+} + Ce^{3+}$

$$Fe^{2+} \rightarrow Fe^{3+} + e^{-} \quad \text{(oxidation half - reaction)}$$
$$\frac{Ce^{4+} + e^{-} \rightarrow Ce^{3+} \qquad \text{(reduction half - reaction)}}{Fe^{2+} + Ce^{4+} \rightarrow Fe^{3+} + Ce^{3+}}$$

- A balanced redox equation has to have charge balance

  - Number of electrons lost in the oxidation half-reaction must be equal to the number of electrons gained in the reduction half-reaction

# CHAPTER 5 - GASES AND GAS LAWS

## Properties of Gases

- Flow freely

- Relatively low densities

  - Lease dense state of matter

- Gases form homogenous mixtures with each other

- Volume of gases change significantly with pressure

  - Volumes of solids and liquids volumes are not greatly affected by pressure

- Volume of gases changes significantly with temperature

  - Under high temperatures gases expand

  - Under low temperatures gases contract

## Pressure

- Force (F) - a function of the mass of an object under acceleration

  - F = Mass x Acceleration

- Pressure (P) - force exerted per unit area of surface by molecules in motion

  - $$P = \frac{Force}{Area} = \frac{ma}{A} = \frac{mg}{A} = \frac{mg}{\frac{v}{h}} = \frac{mgh}{v} = dgh$$

    - A = area

    - m = mass

    - a = acceleration

    - d = density

    - g = acceleration due to gravity = 9.81 m/s$^2$

    - h = height of column (m)

- o Pressure has many units
  - SI unit is Pascal (Pa) = 1 kg•m$^{-1}$•sec$^{-2}$
  - Atmosphere and torr are commonly used
  - 1 atmosphere (atm) = 760 mm Hg = 760 torr

## Gas Laws

*The behavior of gas can be described by pressure (P), temperature (T), volume (V), and molar amount (n). If you hold any of the two variables constant, it allows for determination of a relationship between the other two.*

- Ideal gas – gas that exhibits linear relationships among pressure, temperature, volume, and molar amount
  - o Ideal gases don't actually exist
    - But simple gases behave ideally under normal temperatures and pressures
- Molar gas volume (V$_m$) – volume of one mole of gas
- Volumes of gases are often compared at standard temperature and pressure (STP)
  - o Standard Temperature = 0°C (273 K)
  - o Standard Pressure = 1 atm
  - o V$_m$ at STP = 22.4 L/mol
- Boyles Law
  - o Volume of a sample of gas at a constant temperature is inversely related to the applied pressure
  - o P$_1$V$_1$ = P$_2$V$_2$   or   $\dfrac{P_1}{P_2} = \dfrac{V_1}{V_2}$
- Charles Law
  - o Volume of a sample of gas at constant pressure is directly proportional to the absolute temperature
  - o $\dfrac{V_1}{T_1} = \dfrac{V_2}{T_2}$

- Avogadro's Law
  - Equal volumes of different gases at the same temperature and pressure contain equal number of particles
  - $$\frac{V_1}{n_1} = \frac{V_2}{n_2}$$

- Combined Gas Law
  - For when P, V, and T are changing
    - $$\frac{P_1 V_1}{T_1} = \frac{P_2 V_2}{T_2}$$

- Ideal Gas Law
  - PV = nRT
    - R is the universal gas constant
    - R = 0.082058 L•atm•mol$^{-1}$•K$^{-1}$ = 8.3145 J•mol$^{-1}$•K$^{-1}$
  - As long as you know three of the variables you can manipulate the ideal gas law to solve for the fourth

- Molar Mass from Ideal Gas Law
  - $$M = \frac{mRT}{PV}$$

- Density
  - $$d = \frac{P(MM)}{RT}$$

- Law of Partial Pressures
  - $$P_i = \frac{n_i}{n_{tot}} P_{tot} = x_i P_{tot}$$
    - $x_i = n_i / n_{tot}$

## Kinetic-Molecular Theory of Gases

- A model based on actions of individual atoms
  - Gases consist of particles in constant motion
  - Pressure derived from bombardment with container

- o  Kinetic energy formula: $E_k = \frac{1}{2}mv^2$
- Postulates of Kinetic Theory
    - o  Volume of particles is negligible
    - o  Particles are in constant motion
    - o  No inherent attractive or repulsive forces
    - o  The average kinetic energy of a collection of particles is proportional to the temperature (K)
- Molecular Motion in Gases
    - o  Diffusion – transfer of gas through space or another gas over time
    - o  Effusion – transfer of a gas from a region of high pressure to a region of low pressure
- Grahams Law of Effusion
    - o  $$\frac{effusion\ Rate_A}{effusion\ Rate_B} = \frac{N_A}{N_B}\sqrt{\frac{MM_B}{MM_A}} = \sqrt{\frac{MM_B}{MM_A}}$$
    - o  $$\frac{effusion\ Time_A}{effusion\ Time_B} = \sqrt{\frac{MM_A}{MM_B}}$$

# CHAPTER 6 - THERMOCHEMISTRY

## Thermochemistry

- In chemical reactions whenever matter changes composition, its energy content changes as well

- In some reactions the energy contained in the reactants is higher than the energy contained in the products

    o The excess energy is released as heat

- In other reactions the energy contained in the reactants is lower than the energy contained in the products

    o In these reactions, energy (heat) must be added before the reaction can proceed

- Physical changes also involve a change in energy

- Thermodynamics - science of the relationship between heat and other forms of energy

- Thermochemistry - study of the quantity of heat absorbed or exuded by chemical reactions

## Energy

*Energy is the potential or capacity to do work. Energy is a property of matter and comes in many forms.*

- Forms of energy

    o Radiant energy - electromagnetic radiation

    o Thermal energy - associated with random motion of a molecule or atom

    o Chemical energy - energy stored within the structural limits of a molecule or atom

- Concepts of Energy

    o Kinetic energy ($E_k$) – energy possessed by an object due to its motion

        ▪ $E_k = \frac{1}{2} mv^2$

- o Potential energy ($E_p$) – energy stored in matter because of its position or location

    - ■ $E_p = mgh$

  - o Internal energy ($E_i$ or $U_i$) – energy associated with the random disordered motion of molecules

    - ■ $E_i = E_k + E_p$

- Units of Energy

  - o SI unit of energy is the Joule (J) $= kg \cdot m^2/s^2$

  - o Calorie (cal) - amount of heat required to raise the temperature of one gram of water by one degree Celsius

    - ■ $1 \text{ cal} = 4.181 \text{ J}$

- When reactants interact to form products and the products are allowed to return to the starting temperature, the Internal Energy (E) has changed

  - o $\Delta E$ = change in energy

  - o $\Delta E = E_{final} - E_{initial} = E_{products} - E_{reactants}$

- If energy is lost to the surroundings

  - o $E_{final} < E_{initial}$

  - o $\Delta E < 0$

- If energy is gained from the surroundings

  - o $E_{final} > E_{initial}$

  - o $\Delta E > 0$

## Heat of Reaction

*In chemical reactions, heat is either transferred from the "system" to its "surroundings," or vice versa.*

- Thermodynamic system - quantity of matter or the space under thermodynamic study

- Surroundings - everything in the vicinity of the thermodynamic system that interacts with the system

- Heat (q) - energy that flows into or out of a system because of a difference in temperature between the system and its surroundings
  - Heat flows from a region of higher temperature to a region of lower temperature
    - Once temperatures equalize, heat flow stops
  - When heat is released from the system to the surrounding
    - $q < 0$
    - Reaction is called an **exothermic reaction**
  - When heat is absorbed from the surrounding by the system
    - $q > 0$
    - Reaction is called an **endothermic reaction**
- Heat of reaction - the value of "q" required to return a system to a given temperature when the reaction goes to completion

## Work

*Internal energy is specifically defined as the capacity of a system to do work.*

- Work – energy transferred when an object is moved by a force
  - $w = {}^-P\Delta V$
  - $\Delta E = q_p + w$
  - $\Delta E = q_p + {}^-P\Delta V$
  - $q_p = \Delta E + P\Delta V$
    - $q_p$ - heat absorbed from the surroundings by the system
    - $\Delta E$ - change in internal energy
    - $\Delta V$ - change in volume
    - $P$ – pressure

## Enthalpy and Enthalpy Change

- Enthalpy (H) - an extensive property of a substance that is used to obtain the heat absorbed or exuded in a chemical reaction

    - $H = E + PV$

    - Enthalpy is a state function

        - Property of a system that depends only on its state at the moment and is independent of any history of the system

    - Enthalpy is representative of the heat energy tied up in chemical bonds

- Change in enthalpy ($\Delta H$) - heat added or lost by the system, under constant pressure

    - $\Delta H = \Delta E + P\Delta V$

    - $\Delta H = q_p$

    - Change in enthalpy is also called the enthalpy of reaction

    - $\Delta H_{rxn} = H_{(products)} - H_{(reactants)}$

- If the system has higher enthalpy at the end of the reaction

    - It absorbed heat from the surroundings

    - It is an endothermic reaction

    - $H_{final} > H_{initial}$

    - $\Delta H$ is positive ($+\Delta H$)

- If the system has lower enthalpy at the end of the reaction

    - It exuded heat to the surroundings

    - It is an exothermic reaction

    - $H_{final} < H_{initial}$

    - $\Delta H$ is negative ($-\Delta H$)

## Thermochemical Equation

*Thermochemical equations are chemical reaction equations with the enthalpy of reaction ($\Delta H_{rxn}$) written directly after the equation.*

- Example of a thermochemical equation

  - $2 H_{2 (g)} + O_{2 (g)} \rightarrow 2 H_2O_{(l)}$    $\Delta H_{rxn} = -571.6$ kJ

    - The negative value for $\Delta H_{rxn}$ is telling you that heat is lost to the surrounding

    - Also that the equation is exothermic

- Rules for manipulating thermochemical equations

  - If the thermochemical equation is multiplied by some factor, the value of $\Delta H$ for the new equation is equal to the $\Delta H$ in the original equation multiplied by that factor

  - If the chemical equation is reversed, the sign of $\Delta H$ must be reversed

    - Example, if you were to reverse the direction of the equation mentioned above you would get:

      - $2 H_2O_{(l)} \rightarrow 2 H_{2 (g)} + O_{2 (g)}$    $\Delta H_{rxn} = +571.6$ kJ

## Measuring Heats of Reaction

- Heat capacity – amount of heat required to raise the temperature of an object or substance

  - Varies between substances

- Molar heat capacity (C) – amount of heat required to raise the temperature of **one mole** of a substance by **one degree Celsius**

  - $q = nC\Delta T$

    - $\Delta T = T_{final} - T_{initial}$

- Specific heat capacity (S) – amount of heat required to raise the temperature of **one gram** of a substance by **one degree Celsius**

  - $q = mS\Delta T$

    - $\Delta T = T_{final} - T_{initial}$

    - Units for S: J/g•°C

    - m = grams of a sample

- Hess's Law of Heat Summation

  - For a chemical equation that can be written as the sum of two or more steps, the enthalpy changes for the individual steps can be summed (added) up to determine the enthalpy change for the overall equation

  - For coupled reactions, the individual enthalpy changes can be summed up to determine the enthalpy change for the overall reaction

Hess's Law Example Question: What is the standard enthalpy of reaction for the reduction of iron (II) oxide by carbon monoxide? $FeO_{(s)} + CO_{(g)} \rightarrow Fe_{(s)} + CO_{2(g)}$

- Given Information:

  - Equation 1: $3\ Fe_2O_{3(s)} + CO_{(g)} \rightarrow 2\ Fe_3O_{4(s)} + CO_{2(g)}$     $\Delta H = -48.26\ kJ$

  - Equation 2: $Fe_2O_{3(s)} + 3\ CO_{(g)} \rightarrow 2\ Fe_{(s)} + 3\ CO_{2(g)}$     $\Delta H = -23.44\ kJ$

  - Equation 3: $Fe_3O_{4(s)} + CO_{(g)} \rightarrow 3\ FeO_{(s)} + CO_{2(g)}$     $\Delta H = +21.79\ kJ$

- Changes have to made to the above equations to equal the equation in the question

  - Reverse equation 3 and multiply it by two

    - Puts FeO on the reactant side and moves 2 $Fe_3O_4$ to the products

  - Reverse equation 1

    - Puts $Fe_3O_4$ on opposite side to cancel with the reverse of equation 3

  - Multiply equation 2 by three

    - Gives 3 $Fe_2O_3$ on the reactants that will be used to cancel

- New equations after changes

  - Equation 1: $2\ Fe_3O_{4(s)} + CO_{2(g)} \rightarrow 3\ Fe_2O_{3(s)} + CO_{(g)}$     $\Delta H = +48.26\ kJ$

  - Equation 2: $3\ Fe_2O_{3(s)} + 9\ CO_{(g)} \rightarrow 6\ Fe_{(s)} + 9\ CO_{2(g)}$     $\Delta H = -70.32\ kJ$

  - Equation 3: $6\ FeO_{(s)} + 2\ CO_{2(g)} \rightarrow 2\ Fe_3O_{4(s)} + 2\ CO_{(g)}$     $\Delta H = -43.58\ kJ$

- Summing the three equations gives

  - $6\ FeO_{(s)} + 6\ CO_{(g)} \rightarrow 6\ Fe_{(s)} + 6\ CO_{2(g)}$     $\Delta H = -65.64\ kJ$

- Dividing by six gives the equation in the question and the final answer

  - $FeO_{(s)} + CO_{(g)} \rightarrow Fe_{(s)} + CO_{2(g)}$     $\Delta H = -10.94\ kJ$

## Standard Enthalpies of Formation

- Standard state refers to the standard thermodynamic conditions
    - Pressure - 1 atm (760 mm Hg)
    - Temperature - 25°C (298 K)
- Enthalpy change for a reaction where reactants are in their standard states is called the "Standard Heat of Reaction"
    - $\Delta H°_{rxn}$
- Standard enthalpy of formation of a substance - enthalpy change for the formation of one mole of a substance in its standard state from its component elements in their standard states
    - Standard enthalpy of formation for a "pure" element (C, Fe, O, N, etc.) in its standard state is zero
- Law of Summation of Heats of Formation
    - The standard heat of reaction ($\Delta H°_{rxn}$) is equal to the total formation energy of the products minus the total formation energy of the reactants
        - $\Delta H°_{rxn} = \sum n\Delta H°_f(products) - \sum m\Delta H°_f(reactants)$
        - m and n are coefficients of the substances in the chemical equation

## Summary – Equations

- Kinetic Energy
    - $E_k = \frac{1}{2} mv^2$
- Potential Energy
    - $E_p = mgh$
- Internal Energy
    - $E_i = E_k + E_p$
- Change in Energy
    - $\Delta E = E_{final} - E_{initial} = E_{products} - E_{reactants}$

- Work
  - $w = {}^-P\Delta V$
    - $\Delta E = q_p + w$
    - $\Delta E = q_p + {}^-P\Delta V$
    - $q_p = \Delta E + P\Delta V$
- Enthalpy
  - $H = E + PV$
- Change in Enthalpy
  - $\Delta H = \Delta E + P\Delta V$
  - $\Delta H = q_p$
  - $\Delta H_{rxn} = H_{(products)} - H_{(reactants)}$
- Heat
  - $q = nC\Delta T$
  - $q = mS\Delta T$
  - $-q_{lost}$ (exothermic) $= q_{gained}$(endothermic)
- Law of Summation of Heats of Formation
  - $\Delta H^{\circ}{}_{rxn} = \sum n\Delta H^{\circ}{}_f (products) - \sum m\Delta H^{\circ}{}_f (reactants)$

# CHAPTER 7 – QUANTUM THEORY AND ATOMIC STRUCTURE

## Emission Spectrum

- When elements are burned in a flame and their emissions are passed through a prism only a few color lines are seen

    - These lines are a distinct characteristic of each element

- Atoms emit light of a characteristic wavelengths when they return from an excited state to their ground state

*Emission Spectrum of Hydrogen:* 

## Light

*Light is a form of electromagnetic energy that behaves as both a wave and a particle.*

### Light as a Wave

- Electromagnetic energy travels in rhythmic waves which are disturbances of electric and magnetic fields

- Wavelength (λ) - distance between consecutive crests of electromagnetic waves

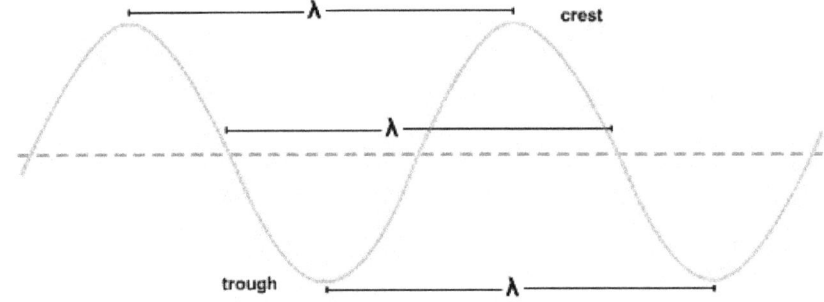

*Wave:*

- Frequency (f) – number of crests of a wave that move past a given point in a given unit of time

    - $f = \lambda / v$

    - $v$ = speed

- Speed of light in a vacuum = 2.998 x 10$^8$ m/sec

- The electromagnetic spectrum encompasses a very wide range of wavelengths from as small as a nanometer to those that are more than a kilometer

## Light as a Particle

- Light also behaves as if it consists of discrete particles or quanta called photons

- Each photon has a fixed quantity of energy

    o E= hc/λ

      ▪ h is the Planck's constant = 6.626 X 10$^{-34}$ joules•sec

      ▪ c is the speed of light in a vacuum = 2.998 X 10$^8$ m/sec

    o Shorter λ = higher energy

## Wave-Particle Duality

- All matter and energy exhibit both wave and particle-like properties

- This duality is seen in:

    o Macroscopic objects

    o Microscopic objects (e.g. atoms and molecules)

    o Quantum objects (e.g., protons, neutrons, quarks, mesons)

- To explain Wave-Particles Duality, physicists focused on 3 phenomena
  - Black Body Radiation
  - Photoelectric Effect
  - Atomic Line Spectra

## Black Body Radiation

*As the temperature of an object changes, its temperature is directly related to the wavelengths of light that it emits. This is a characteristic of the idealized "black body."*

- Max Planck developed a mathematical model to reproduce the spectrum of light emitted by glowing objects
  - The model was developed under the assumption that a vibrating (oscillating) atom can only emit or absorb certain quantities of energy
- Planck's Model
  - $E = nh\nu$
    - E - energy of radiation
    - $\nu$ – frequency
    - n – quantum number (1, 2, 3...)
    - h – Planck's Constant = $6.626 \times 10^{-34}$ J•s
  - Quanta – packets of energy that can be emitted or absorbed
  - Atoms change energy states when they emit or absorb one or more quanta
  - The model views emitted energy as waves

## Photoelectric Effect

*The photoelectric effect is the observation that many metals emit electrons when light shines upon them. Electrons emitted in this manner are called photoelectrons.*

- Electrons are ejected from the surface of the metal only when the frequency exceeds a certain threshold
  - The threshold is dependent on the characteristic of the metal

- Einstein reasoned that atoms change energy states when they emit or absorb a quantum of light energy, which he called a photon

  o He defined a photon as a particle of electromagnetic energy

  o $E_{photon} = \Delta E_{atom} = h\nu = \frac{hc}{\lambda}$

- According to the photoelectric effect

  o Electrons exist in different energy states

  o A photon whose frequency is greater than or equal to the energy state of the electron will be absorbed

    ▪ If the frequency is less than the electron energy level, the photon is not absorbed

  o The electron moves to a higher energy state and is ejected from the surface of the metal when it absorbs a photon

  o Electrons are attracted to the positive anode of a battery which causes a flow of current

## Atomic Line Spectra

- When light from an "excited" atom passes through a prism, it does not form a continuous spectrum

  o Instead, it produces a series of colored lines called "line spectra" that are separated by black spaces

  o Wavelengths of the lines are a characteristic of the elements producing them

    ▪ Different elements have different line spectra

- Example of Hydrogen

  o The spectra lines of Hydrogen occur in several series

    ▪ Each series is represented by a positive integer, n

Line Spectra of Hydrogen:

Ultraviolet series — Visible series — Infrared series

0   200   400   600   800   1000   1200   1400   1600   1800   2000 nm

- Ultraviolet series, n =1

- Visible series, n = 2

- Infrared series, n = 3

o The visible spectrum of Hydrogen can be reproduced by a Rydberg Equation

$$\frac{1}{\lambda} = R\left(\frac{1}{n_1^2} - \frac{1}{n_2^2}\right) = 1.096776 \; x \; 10^7 \; m^{-1} \left(\frac{1}{n_1^2} - \frac{1}{n_2^2}\right)$$

- R – Rydberg Constant = 1.096776 x $10^7$ m$^{-1}$

- Λ – wavelength of the spectra line

- $n_1$ and $n_2$ are positive integers where $n_2 > n_1$

## Bohr Theory

*Bohr's theory of an atom was an early model of atomic structure in which electrons travel around the nucleus in a number of discrete stable orbits determined by quantum conditions.*

- Bohr came up with 2 postulates to account for how electrons lose energy yet remain in orbit

  o A electron has specific energy levels in an atom

  o A electron in an atom changes energy levels by undergoing a transition from one energy level to another

- Bohr came up with a formula for the energy levels of the electron in the hydrogen atom

  - $E = -2.18 \times 10^{-18} \text{ J} \left( \dfrac{Z^2}{n^2} \right)$

    - $n$ – principal quantum numbers = 1, 2, 3...

    - $Z$ – nuclear charge

      - $Z = 1$ for Hydrogen

- When an electron undergoes a transition from a higher energy level ($n_i$) to a lower one ($n_f$) the energy is emitted as a photon

  - $E = h\nu = E_f - E_i = \Delta E$

    - $E$ – energy of emitted photon

    - $E_i = -\dfrac{2.18 \times 10^{-18} \text{ J}}{n_i^2}$

    - $E_f = -\dfrac{2.18 \times 10^{-18} \text{ J}}{n_i^2}$

  - $\Delta E = h\nu = -2.18 \times 10^{-18} \text{ J} \left( \dfrac{1}{n_f^2} - \dfrac{1}{n_i^2} \right)$

## Quantum Mechanics

*Quantum mechanics is the branch of physics that attempts to mathematically describe the wave properties of submicroscopic particles.*

- de Broglie relation

  - $\lambda = h/mv$

    - Matter has wave-like properties but these properties are not commonly observed

    - de Broglie relation shows that the wavelength of objects is so incredibly small that these waves cannot be detected

- Heisenberg's Uncertainty Principle

  - States that you can never know the exact position and the exact speed of an object with high precision

    - Essentially, there is an uncertainty

- The principle is also a relationship that states that the product of the uncertainty in position ($\Delta x$) and the uncertainty in momentum ($m\Delta v_x$) of a particle can be no smaller than what is predicted by the equation:

  - $(\Delta x)(m\Delta v_x) \geq \dfrac{h}{4\pi}$

    - $\dfrac{h}{4\pi} = 5.28 \times 10^{-35}$ J•s

  - When m is large (represents a large object) the uncertainty is very small but for subatomic particles like electron there is a high level of uncertainty

    - This uncertainty is why we can't define the exact orbit of an electron

- Schrodinger developed a quantum mechanical model of the hydrogen atom. According to the model:

  - An atom has specific allowed quantities of energy

  - An electron's behavior is wave-like

    - However, its exact location is impossible to know

  - The electron's Matter-Wave occupies a 3-D space near the nucleus

    - The Matter-Wave experiences continuous and varying influence from the nuclear charge

- Schrodinger Equation

  - $H^{(op)}\Psi = E\Psi$

    - E – energy of the atom

    - $\Psi$ – wave function

    - $H^{(op)}$ – Hamiltonian Operator

**Quantum Numbers and Atomic Orbitals**

- Quantum mechanics describes each electron by 4 quantum numbers

  - Principal Quantum Number (n)

  - Angular Momentum Quantum Number (l)

  - Magnetic Quantum Number ($m_\ell$)

- o Spin Quantum Number ($m_s$)
  - ▪ n, $\ell$, $m_\ell$ define the wave function of the electron's atomic orbital
  - ▪ $m_s$ refers to the spin orientation of the 2 electrons that occupy an atomic orbital

- Principal Quantum Number (n) represents the "shell number" in which an electron is located
  - o Represents the relative size of the orbital
  - o Defines the principal energy of the electron
  - o Smaller "n" represents a smaller orbital and a lower energy of the electron
  - o n can take any positive value

- Angular Momentum Quantum Number ($\ell$) represents the "sub shells" within a given shell
  - o Each main "shell" is designated by a quantum number "n"
    - ▪ It is further subdivided into: $\ell$ = n - 1 "sub shells"
    - ▪ "$\ell$" can have any integer value from 0 to n - 1
    - ▪ "$\ell$" values correspond to the s, p, d, f designations used in the electronic configuration of the elements
      - s has an $\ell$ value of 0
      - p has an $\ell$ value of 1
      - d has an $\ell$ value of 2
      - f has an $\ell$ value of 3

- Magnetic Quantum Number ($m_\ell$) defines atomic orbitals within a given sub-shell
  - o Each value of the angular momentum number ($\ell$) determines the number of atomic orbitals
  - o For any given value of "$\ell$," $m_l$ can take any integer value from $-\ell$ to $+\ell$
    - ▪ $m_\ell = -\ell$ to $+\ell$
  - o Each orbital has a different shape and orientation (x, y, z) in space
  - o Each orbital within a given angular momentum number sub shell ($\ell$) has the same energy

- Spin Quantum Number ($m_s$) represents the two possible spin orientations of an electron residing within a given atomic orbital
    - Each atomic orbital holds only two electrons
    - Each electron has a "spin" orientation value
        - These values must be opposite of one another
    - Possible values of $m_s$ are: $+\frac{1}{2}$ and $-\frac{1}{2}$

*Summary of Quantum Numbers:*

| Name | Symbol | Permitted Values | Property |
|---|---|---|---|
| Principal | $n$ | positive integers (1, 2, 3...) | orbital energy (size) |
| Angular Momentum | $\ell$ | integers from 0 to n-1 | orbital shape The $\ell$ values 0, 1, 2, and 3 correspond to s, p, d, and f orbitals, respectively |
| Magnetic | $m_\ell$ | integers from $-\ell$ to 0 to $+\ell$ | orbital (x,y,z) orientation |
| Spin | $m_s$ | $+\frac{1}{2}$ or $-\frac{1}{2}$ | e⁻ spin orientation |

- Shapes of orbitals
    - s (n=1) sub shell orbital
        - Only one orbital
        - Holds 2 electrons
        - Spherical shape
    - p (n=2) sub shell orbitals
        - Three orbitals
        - Holds 6 electrons
        - Dumbbell shape
    - d (n=3) sub shell orbital
        - Five orbitals
        - Holds 10 electrons
        - Pear-shaped lobes, and dumbbell shapes

- f (n=4) sub shell orbitals
  - Seven orbitals
  - Holds 14 electrons

# CHAPTER 8 – ELECTRON CONFIGURATION AND PERIODIC PROPERTIES

## Brief Review to Understand Electron Configuration

- The quantum numbers n, $\ell$, $m_\ell$ define an orbital

    - A orbital contains a maximum of 2 electrons

        - Each electron has a different spin (+½ or -½ )

- Orbital diagrams are notations used to show how the orbitals of a sub shell are occupied by electrons

    - Each group of orbitals is labeled by its sub shell notation (s, p, d, f)

    - Electrons are represented by arrows

        - Up ↑ for $m_s$ = +½

        - Down ↓ for $m_s$ = -1/2

*Example of an Orbital Diagram:*

- Pauli Exclusion Principle

    o No two electrons in an atom can have the same four quantum numbers

- Summary table for the maximum number of electrons in each sub shell

| Sub Shell | Number of Orbitals | Values for ($-\ell$ to $+\ell$) | Maximum Number of Electrons |
|---|---|---|---|
| s ($\ell =0$) | 1 | (0) | 2 |
| p ($\ell =1$) | 3 | (-1, 0, +1) | 6 |
| d ($\ell =2$) | 5 | (-2, -1, 0, +1, +2) | 10 |
| f ($\ell =3$) | 7 | (-3, -2, -1, 0, +1, +2, +3) | 14 |

# Electron Configuration

*The electron configuration of an atom refers to the particular distribution of electrons among the available sub shell in that atom.*

- Electronic configuration notation lists subshell symbols (s, p, d, f) sequentially with a superscript to indicate the number of electrons in that subshell

    o Ex. Fluorine

        - Atomic Number: 9

        - Number of electrons in a neutral Fluorine atom: 9

            - Number of electrons for a neutral atom is the same as the atomic number

        - 2 electrons in the "1s" sub shell

        - 2 electrons in the "2s" sub shell

        - 5 electrons in the "2p" sub shell

        - Electron Configuration: $1s^22s^22p^5$

- Configurations can become quite complex as atomic number increases

    o To remedy this, a condensed form of the configuration is often used which utilizes electron configurations of noble gases

        - Noble gases have the maximum number of electrons possible in their outer shell

            - Makes them very unreactive

        - The noble gases are: Helium, Neon, Argon, Krypton, Xenon, and Radon

| Element | Noble Gas? | Full Electronic Configuration | Condensed Electronic Configuration | Total Number of Electrons |
|---------|------------|-------------------------------|-------------------------------------|---------------------------|
| Neon | Yes | $1s^22s^22p^6$ | [Ne] | 10 |
| Argon | Yes | $1s^22s^22p^63s^23p^6$ | [Ar] | 18 |
| Krypton | Yes | $1s^22s^22p^63s^23p^63d^{10}4s^24p^6$ | [Kr] | 36 |
| Beryllium | No | $1s^22s^2$ | [He] $2s^2$ | 4 |
| Magnesium | No | $1s^22s^22p^63s^2$ | [Ne] $3s^2$ | 12 |
| Calcium | No | $1s^22s^22p^63s^23p^64s^2$ | [Ar] $4s^2$ | 20 |

- o [X] represents the electron configuration of the nearest noble gas that appears before the element of interest on the periodic table
- Keep in mind that you have to adjust the number of electrons and thus the electron configuration for cations and anions of an element

## Nuclear Charge, Shielding Effect, and Orbital Shape

### Nuclear Charge

- In an atom there are 2 counteracting forces:
  - o In the nucleus
    - Positive protons pull (attract) the negatively charged electrons
  - o Outside the nucleus
    - Negatively charged electrons are repelling each other
- Higher nuclear charge (Z) lowers orbital energy by increasing the amount of proton-electron attractions
  - o Lowering orbital energy makes it more difficult to remove the electron from orbit

### Shielding Effect

- The repulsion electrons experience from other electrons shields (counteracts) the attractive force of the protons in the nucleus

- Shielding lowers the full nuclear charge to an "effective nuclear charge" ($Z_{eff}$)
  - Lowering the effective nuclear charge makes it easier to remove an electron
    - It takes less energy to remove an electron from Helium (He) than from $He^+$
      - Since the second electron in He repels the first electron and effectively shields the first electron from the full nuclear charge

**Effects of Orbital Shape**

- Shape of an atomic orbital affects how close an electron comes to the nucleus (i.e. the level of penetration)
- Penetration and shielding cause the energy level (n) to be split into sublevels of differing energy
  - This is represented by the various values of the magnetic quantum number ($\ell$)
    - The lower the value of the magnetic quantum number, the greater the electron penetration
  - Order of Sublevel Energies
    - s ($\ell = 0$) < p($\ell = 1$) < d($\ell = 2$) < f($\ell = 3$)

## Aufbau Principle

*The Aufbau principle states that electrons orbiting one or more atoms fill the lowest available energy levels before filling higher levels. For example, an electron has to fill the "1s" before "2s."*

- Filling orbitals of the lowest energy first gives the lowest total energy of the atom
  - The ground state of the atom

- Order in which the possible sub-shells fill

  o 1s, 2s, 2p, 3s, 3p, 4s, 3d, 4p, 5s, 4d, 5p, 6s, 4f, 5d, 6p, 7s, 5f

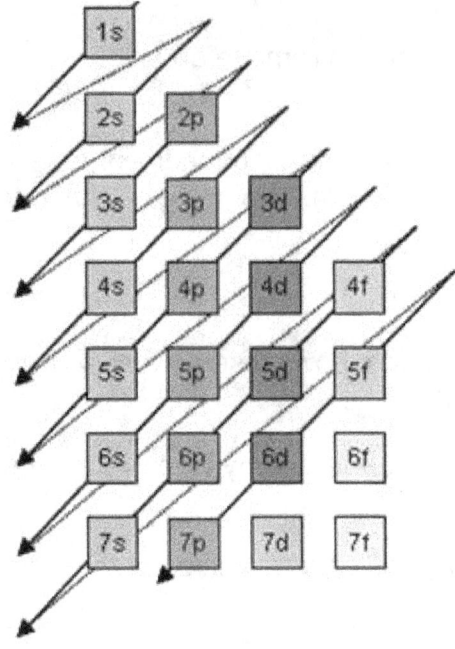

*Order of Filling Subshells:*

  o Order in which subshells are filled is in the order in which the diagonal lines go through the subshells

    ▪ Note: 4s subshell is filled before the 3d subshell because 4s electrons are at a lower energy level than the 3d electrons

    ▪ Lower energy levels are always filled first

- Configuration associated with the lowest energy level of an atom is the ground state

  o Any other configurations correspond to the excited states

## Hund's Rule

*Hund's rule states that the "ground state" (i.e. the lowest energy configuration) of electrons in a sub-shell is attained from putting electrons into separate orbitals of a sub shell before pairing the electrons.*

### Example: Oxygen

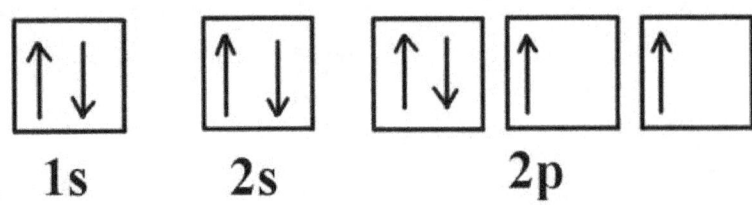

*Oxygen Orbitals:*    **1s**        **2s**              **2p**

- Note: Two of the electrons in the 2p orbitals appear singly instead of being paired together
  - According to Hund's Rule, electrons occupy separate orbitals rather than pairing when possible

## Configurations and the Periodic Table

- Electrons in an atom's outermost shell are called valence electrons

  - Valence electrons are the electrons primarily involved in chemical reactions

  - Elements within a group have the same valence shell configuration

    - This is why groups of elements share chemical properties

- Noble gases have filled outer shells

  - Very high ionization energies

  - Positive (endothermic) electron affinities

  - Do not readily form ions or react

    - Inert for the most part

  - Very stable

    - Other elements try to attain noble gas configuration (filled outer shells) for stability

- Elements in Groups 1A and 2A readily form cations by losing electrons

  - They only have 1 or 2 electrons in their outer shells respectively, losing their valence electrons allows these elements to conform to noble gas configuration

- Elements in Groups 6A and 7A readily form anions by gaining electrons

  - They fill their outer shells and conform to a noble gas configuration

  - They are isoelectronic with the nearest noble gas configuration

    - Isoelectronic – have the same number of electrons or the same electronic structure

- Large metals from Groups 3A, 4A, and 5A form cations through a different process than the smaller 1A and 2A elements

  - They have to lose large amounts of electrons in order to attain noble gas configuration

  - Example: Tin (Sn), would have to lose two electrons from 5p, ten from 4d, two from 5s in order to be isoelectronic with Krypton

    - For Tin it is easier to gain stability by reaching a noble gas-like configuration by having empty 5s & 5p sublevels and a filled inner 4d subshell configuration

- Many times it is only important to know the valence electron number for an atom

  - On the periodic table it is easy to determine this: the number of valence electrons for an atom matches its group number

    - Example: Group 7A elements have 7 valence electrons

## Periodic Properties

*Periodic Law states that when elements are arranged by atomic number, their physical and chemical properties vary across the periodic table row.*

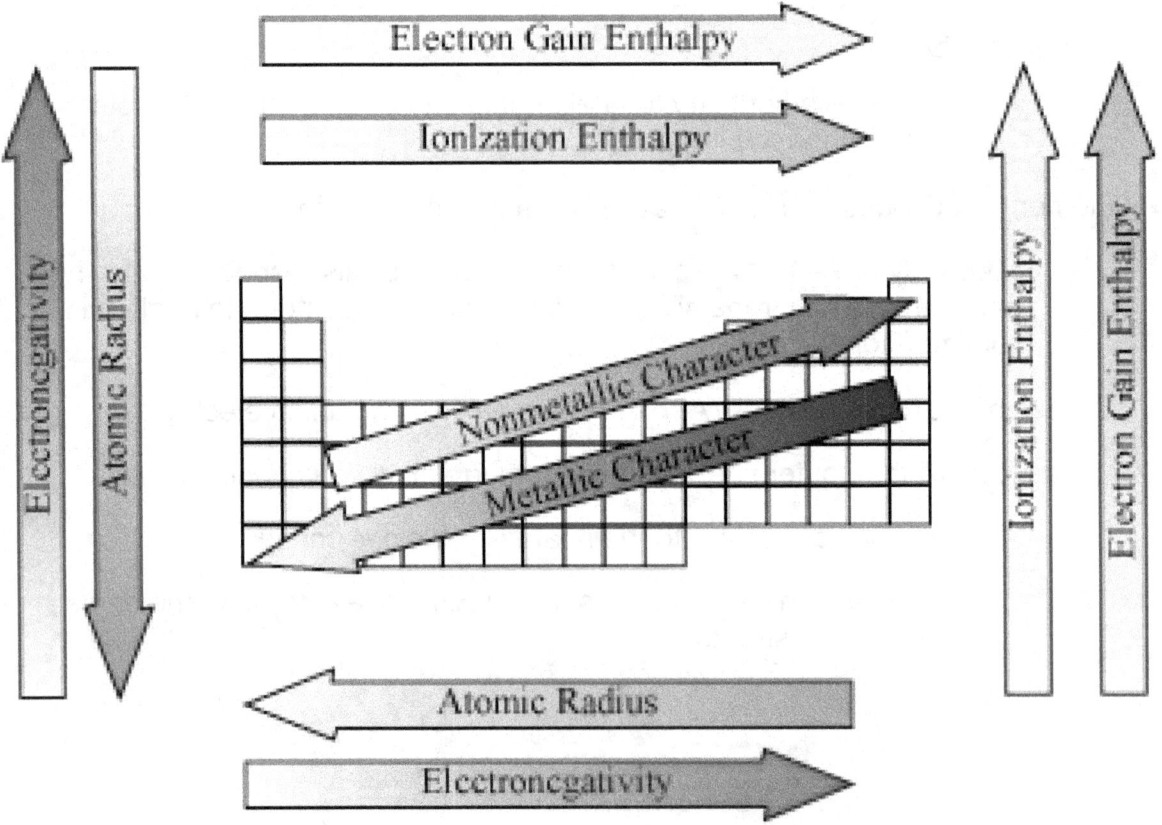

**Atomic Size**

- Two factors that affect size of an atom

  - Larger the principal quantum number (n), the larger the size of the orbit

  - Effective nuclear charge

    - Positive charge an electron experiences from the nucleus minus any shielding effects

- Atomic radius tends to **decrease** with increasing atomic number across a period

  - There is a higher effective nuclear charge

    - Greater attractive force in the nucleus from the higher number of protons

- Atomic radius tends to **increase** with increasing atomic number within a group

  - More energy levels

    - Each subsequent energy level is further from the nucleus than the last

**Ionic Size**

- Cations are **smaller** than their neutral atom counterparts

  - Electrons are removed

    - Results in a decrease in electron repulsion

      - Allows nuclear charge to pull electrons closer

- Anions are **bigger** than their neutral atom counterparts

  - Electrons are added

    - Results in an increase in electron repulsion

      - Occupy more space

- Ionic size **increases** down a group

  - More energy levels

- Ionic size is slightly complicated across a period

  - Decreases among cations

  - Increases dramatically with the first anion

- Decreases within anions

## Ionization Energy (IE)

- First ionization energy - the minimal energy needed to remove one of the outermost electrons from a neutral atom

  - Successive electron removal is called second ionization energy, third ionization energy, etc.

  - Successive ionization energies increase because each electron that is pulled away creates a larger positive charge (i.e. a higher effective nuclear charge)

- Ionization energies tend **increase** with atomic number within a period

  - More difficult to remove an electron that is closer to the nucleus

    - Remember: There is a higher effective nuclear charge across a period

- Ionization energies tend to **decrease** down a group

## Electron Affinity

*This is not the same as electronegativity!*

- More negative electron affinity value express that a more stable negative ion is formed

  - Negative values indicate that energy is released when the anion for that element forms

- General trend is that values become more negative from lower left to upper right

  - Highest electron affinities occur for F and Cl

## Electronegativity

- Electronegativity is the measure of an atom's ability of to draw bonding electrons to itself in a molecule

- Electronegativity tends to increase from the lower-left corner to the upper-right corner of the periodic table

## Metals, Non-Metals, and Metalloids

**Metals**

- Typical Properties
    - Shiny solids
    - High melting points
    - Good thermal & electrical conductors
    - Malleable
    - Lose their electrons to non-metals

**Non-Metals**

- Typical Properties
    - Not shiny
    - Low melting points
    - Poor thermal & electrical conductors
    - Crumbly solids or gases
    - Gain electrons from metals

**Metalloids (Semi-metals)**

- Elements that exhibits external characteristics of a metal, but behaves chemically as a nonmetal
- Typical Properties
    - Make good semiconductors
    - Intermediate electronegativity values, conductivity, and ionization energies
    - Reactivity depends on the element with which they are reacting
    - Do not form multiple bonds
        - Compounds with metalloids often have incomplete octet around the central atom

**Metallic Behavior Trends**

- Metallic behavior decreases across (left to right) a period and increases down a group

# Chapter 9 – Chemical Bonding

## Ionic Bonds

*Ionic bonds are chemical bonds formed by the electrostatic attraction between positive and negative ions.*

- Ionic bonds involve the transfer of electrons from one atom to another

  - Usually the transfer is from a metal from Group IA or IIA to a nonmetal from Group 7A or the top of Group 6A

- Number of electrons lost or gained by an atom is determined by its need to be isoelectronic with its nearest noble gas

  - Noble gas configurations are extremely stable

- Ionic bonds result in the formation of ions that are electrostatically attracted to one another

  - Anion – negatively charged ion

    - Size of an anion is larger than the original size of the neutral atom

  - Cation – positively charged ion

    - Size of a cation is smaller than the original size of the neutral atom

### Properties of Ionic Compounds

- Hard

  - Don't dent

- Rigid

  - Don't bend

- Brittle

  - Crack but don't deform

- The above properties are a result of the powerful attractive forces holding ions together

  - Moving ions out of position requires large amounts of energy to overcome the attractive forces

## Covalent Bonds

- Two nonmetals often form covalent bonds

  - Share electrons since they have similar attractions for them

- Each nonmetal holds tightly to its own electrons

  - Shared electron pair spend most of their time between the two atoms

    - Electron pairs being shared are said to be localized

- How covalent bonds form

  - Distance between two nuclei decreases

    - Each starts to attract the other's electron(s)

    - Causes a decrease in potential energy

  - Atoms draw closer and closer together

    - Energy becomes progressively lower

  - Attractions increase but so does repulsions between electrons

  - At a particular internuclear distance, maximum attraction is achieved

    - Balance between nucleus-electron attractions

    - Balance between electron-electron and nucleus-nucleus repulsions

- Two sets of forces are involved within covalent compounds

  - Strong covalent bonding forces hold atoms together within the molecule

  - Weak intermolecular forces hold separate molecules near each other

    - Weak intermolecular forces between molecules are responsible for the observed physical properties of these molecules

## Types of Covalent Bonds

- Coordinate covalent bond – covalent bond in which both of the shared electrons are donated by a single atom

- Double bond – sharing of two pairs of electrons between atoms

- Triple bonds – sharing of three pairs of electrons between atoms

- Polar covalent bond – covalent bond where the bonding electron spend more time closer to one of the atoms involved in the bonding

- Nonpolar covalent bond – covalent bond where the bonding electrons are shared equally

## Bonding Pairs and Lone Pairs

- Shared electrons are considered as belonging entirely to each atom in a covalent bond
    - Shared electron pair simultaneously fills the outer level of both atoms
- An outer-level electron pair that is not involved in bonding is a "lone pair"

## Metal-Metal Bonding

- Metals have low ionization energies
    - Lose electrons easily
    - Do not gain them readily
- Valence electrons are evenly distributed around metal-ion cores
    - Metal-ion cores consist of the nucleus plus the inner electrons
- Valence electrons are delocalized
    - Move freely throughout the metal

## Predicting Ionic and Covalent Bonding

- Non-polar covalent bond
    - Typically electronegativity difference between the two atoms has to be less than 0.5 for non-polar bonds
- Polar covalent bonds
    - Electronegativity between the two atoms is different by a greater degree than 0.5 but less than 2.0
- Ionic bonds
    - Typically, difference in electronegativity is more than 2.0

# Bond Length, Bond Order, and Bond Energy

- Bond order - measure of the number of bonding electron pairs between atoms

  - Single bonds have a bond order of 1

  - Double bonds have a bond order of 2

  - Triple bonds have a bond order of 3

  - Fractional bond orders are possible in molecules and ions that have resonance structures

- Bond length (aka bond distance) – distance between the nuclei in a bond

  - Bond length depends on bonds order

  - Increase in bond order means a shorter and stronger bond

- Bond Energy (BE) – measure of the amount of energy required to break apart one mole of covalently bonded gases

  - Quantity of heat absorbed to break reactant bonds is denoted $\Delta H°_{reactants}$

  - Quantity of heat released to form product bonds is denoted $\Delta H°_{products}$

  - Exothermic reactions

    - $\Delta H°_{rxn}$ is negative

  - Endothermic reactions

    - $\Delta H°_{rxn}$ is positive

  - $\Delta H°_{rxn} = \sum BE_{reactant\ bonds\ broken} - \sum BE_{product\ bonds\ formed}$

# CHAPTER 10 – GEOMETRY OF MOLECULES

**Lewis Structures**

*Lewis Structure of Carbon:*

- Lewis dot symbols represent electrons in the valence shell of an atom or ion as dots placed around the letter symbol of the element

    o Bonding electron pairs are represented by either two dots or a dash

*Lewis Electron-dot Formula Example:*

$$H \cdot + :\overset{..}{\underset{..}{Cl}} \cdot \longrightarrow H - \overset{..}{\underset{..}{Cl}}: \quad or \quad H:\overset{..}{\underset{..}{Cl}}:$$

- Rules for Forming Lewis Structures

    o Calculate the number of valence electrons for the molecule

        ▪ Group # for each atom (1-8)

            • Gives valence electron number for each atom

            • Add all numbers up

        ▪ Add the charge of any anions

            • Example: an anion with a -2 charge has 2 extra electrons, you would add 2 to the total count

        ▪ Subtract the charge of any cations

            • Example: a cation with a +3 charge lacks 3 electrons, you would subtract 3 from the total count

    o Place the atom with the lowest group number and lowest electronegativity as the central atom

    o Arrange the other elements around the central atom

    o Distribute electrons to atoms surrounding the central atom to satisfy the octet rule for each atom o Distribute the remaining electrons as pairs to the central atom

o If the central atom is deficient in electrons, complete the octet for it by forming double bonds or possibly a triple bond

## Octet Rule

*The octet rule states that the tendency of atoms in a molecule is to have eight electrons in their outer shell.*

- There are exceptions to this rule where the central atom may have more than eight electrons

- Generally, a nonmetal in the third period or higher can accommodate as many as twelve electrons, if it is the central atom

  o These elements have unfilled "d" subshells that can be used for bonding

## Resonance (Delocalized Bonding)

- Structures of some molecules can be represented by more than one Lewis dot formula

*Example of Resonance Structures:*

- Delocalized bonding is where a bonding pair of electrons is spread over a number of atoms

- Some resonance structures are more likely to occur than others

  o Formal charges can help discern which structure is most likely

**Formal Charge**

- An atom's formal charge is:
    - Total number of valence electrons
    - Minus all unshared electron
    - Minus ½ of its shared electrons
- Formal charges have to sum to the actual charge of the species
    - 0 charge for a molecule
    - Ionic charge for an ion
- Lewis structures with the smallest formal charge are the most likely to occur

**Formal Charge vs. Oxidation Number**

- Formal charges are used to examine resonance hybrid structures
    - Oxidation numbers are used to monitor redox reactions
- Formal charge
    - Bonding electrons are assigned equally to the atoms
        - Each atom has half the electrons making up the bond
    - Formal Charge = valence $e^-$ – (unbonded $e^-$ + ½ bonding $e^-$)
- Oxidation Number
    - Bonding electrons are transferred completely to the more electronegative atom
    - Oxidation Number = valence $e^-$ – (unbonded $e^-$ + bonding $e^-$)

## Valence-Shell Electron Pair Repulsion Model (VSEPR)

*VSEPR predicts the shapes of molecules and ions by assuming that the valence shell electron pairs are arranged as far from one another as possible.*

- Molecular geometry – 3-D arrangement of atoms that constitute a molecule
    - Shape of a molecule is determined by the positions of atomic nuclei relative to each other

- Rules to help discern electron pair arrangements

  o Select the central atom

    ▪ Place atom with the lowest group number in the center

      • If atoms share the same group number place the atom with the higher period number in the center

  o Draw the Lewis structure

  o Determine the number of bonding electron pairs around the central atom

  o Determine the number of non-bonding electron pairs

    ▪ Multiple bonds are counted as a single electron pair

  o Arrange the electron pairs as far apart as possible

    ▪ Minimizes electron repulsions

  o Add the number of bonding and lone pairs

  o From that number and the number of lone pairs you can use the chart that follows to determine the geometry

    ▪ For example: If you were given a molecule where the central atom had 2 bonding pairs and 1 non-bonding pair (total number =3), the molecular shape would be *bent/angular*

## VSEPR Geometries

| Steric No. | Basic Geometry 0 lone pair | 1 lone pair | 2 lone pairs | 3 lone pairs | 4 lone pairs |
|---|---|---|---|---|---|
| 2 | Linear (180°) | | | | |
| 3 | Trigonal Planar (120°) | Bent or Angular (< 120°) | | | |
| 4 | Tetrahedral (109°) | Trigonal Pyramid (< 109°) | Bent or Angular (<< 109°) | | |
| 5 | Trigonal Bipyramid (120°, 90°) | Sawhorse or Seesaw (< 120°, < 90°) | T-shape (< 90°) | Linear (180°) | |
| 6 | Octahedral (90°) | Square Pyramid (< 90°) | Square Planar (90°) | T-shape (< 90°) | Linear (180°) |

Example Question: Determine the molecular shape of $CO_2$ according to VSEPR theory.

$CO_2$: O=C=O

- Central atom: C
- C has 2 bonding pairs, 0 non-bonding pairs
  - Remember that double bonds are counted as a single electron pair
- According to VSEPR model the bonds are arranged linearly
  - Bond angle = 180°
- Molecular shape is *linear*

Example Question: Determine the molecular shape of $COCl_2$ according to VSEPR theory.

*Example of $COCl_2$:*

- Central atom: C

- C has 3 bonding pairs, 0 non-bonding pairs

- According to VSEPR model the three groups of electron pairs are arranged in a trigonal plane

  o Bond angle = 120°

- Molecular shape is *trigonal planar*

**Effect of Lone Pairs**

- Lone pairs are less confined because they are held by a single nucleus

  o Allows them to exert a stronger repulsive force than a bonding pair

    ▪ Results in a decrease in the angle between bonding pairs

## Dipole Moment

- Dipole moment (μ) - measure of the degree of charge separation (molecular polarity) in a molecule

  o $\mu = Q*r$

    ▪ Q = Charge

    ▪ r = distance between the charges

- Polarity of individual bonds within a molecule can be viewed as vector quantities

  o Molecules that are perfectly symmetric have a zero dipole moment

    ▪ These molecules are considered nonpolar

  o Molecules that exhibit any asymmetry in the arrangement of electron pairs will have some dipole moment

    ▪ These molecules are considered polar

**Net moment =** ↑

*Example of H₂O Polarity Vectors:*

*Molecular Geometry and Dipole Moment Relatedness*

| Molecular Geometry | Dipole Moment |
|---|---|
| Linear | Can Be nonzero |
| Linear | Zero |
| Trigonal Planar | Zero |
| Trigonal Planar Bent | Can Be nonzero |
| Tetrahedral | Zero |
| Tetrahedral Trigonal Pyramidal | Can Be nonzero |
| Tetrahedral Bent | Can Be nonzero |
| Trigonal Bipyramidal | Zero |
| Trigonal Bipyramidal SeeSaw | Can Be nonzero |
| Trigonal Bipyramidal T-Shaped | Can Be nonzero |
| Trigonal Bipyramidal Linear | Can Be nonzero |
| Octahedral | Zero |
| Octahedral Square Pyramidal | Can Be nonzero |
| Octahedral Square Planar | Zero |

# CHAPTER 11 – BONDING THEORIES

## Valence Bond Theory

*Valence bond theory is an attempt to explain the covalent bond from a quantum mechanical view.*

- Orbitals (s, p, d, f) of the same type have the same energy

- According to the theory, a bond forms when two atomic orbitals overlap

    o Space formed by overlapping orbitals has a capacity for two electrons

        ▪ Must have opposite spins (+½ and -½)

    o Each orbital forming the bond has at least one unfilled slot to accommodate the electron being shared

    o Bond strength depends on the attraction of the nuclei for the shared electrons

        ▪ Greater orbital overlap = stronger bond

    o Overlap depends on the shapes and direction of the orbitals

## Hybrid Orbitals

- Quantum mechanical calculations show that if specific combinations of orbitals are mixed, "new" atomic orbitals are formed

    o These new orbitals are called hybrid orbitals

- Types of hybrid orbitals

  - Each type has a unique geometric arrangement

| Hybrid Orbitals (Hybridization) | Geometric Arrangements | Number of Hybrid Orbitals Formed by Central Atom | Example |
|---|---|---|---|
| sp | Linear | 2 | Be in $BeF_2$ |
| $sp^2$ | Trigonal planar | 3 | B in $BF_3$ |
| $sp^3$ | Tetrahedral | 4 | C in $CH_4$ |
| $sp^3d$ | Trigonal bipyramidal | 5 | P in $PCl_5$ |
| $sp^3d^2$ | Octahedral | 6 | S in $SF_6$ |

- Hybrid orbitals are used to describe bonding that is obtained by taking combinations of atomic orbitals of an isolated atom

- Steps for determining bonding description

  - Write the Lewis dot formula for the molecule

  - Then use the VSEPR theory to determine the arrangement of electron pairs around the central atom

  - From the geometric arrangement, determine the hybridization type

  - Assign valence electrons to the hybrid orbitals of the central atom one at a time

    - Pair only when necessary

  - Form bonds to the central atom by overlapping singly occupied orbitals of other atoms with the singly occupied hybrid orbitals of the central atom

**Multiple Bonds**

- Orbitals can overlap two ways

  - Side to side

  - End to end

- Two types of covalent bonds

    o Sigma bonds (C-C)

        ▪ Formed from an overlap of one end of the orbital to the end of another orbital

    o pi bonds (C=C)

        ▪ Formed when orbitals overlap side to side

        ▪ Creates two regions of electron density

            • One above and one below

- Double bonds always consist of one sigma bond and one pi bond

## Molecular Orbital Theory

- As atoms approach each other and their atomic orbitals overlap, molecular orbitals are formed

    o Only outer (valence) atomic orbitals interact enough to form molecular orbitals

- Combining atomic orbitals to form molecular orbitals involves adding or subtracting atomic wave functions

- Adding wave functions

    o Forms a bonding ($\sigma$) molecular orbital

    o Electron charge between nuclei is dispersed over a larger area than in atomic orbitals

    o Molecular orbitals have lower energy than atomic orbitals

        ▪ Reduction in electron repulsion

    o Bonding molecular orbital is more stable than atomic orbital

- Subtracting Wave Functions

    o Forms a antibonding ($\sigma^*$) molecular orbital

    o Electrons do not shield one nuclei from the other

        ▪ Results in increased nucleus-nucleus repulsion

    o Antibonding molecular orbitals have a higher energy than the corresponding atom orbitals

- ○ When the antibonding orbital is occupied, the molecule is less stable than when the orbital is not occupied

*Molecular Orbitals of H₂:*

# CONCLUDING REMARKS

I hope this book has provided you tremendous value for your money and has helped you do better on your exams! If it has done both of these things, I have achieved my purpose in making this guide.

Furthermore, my goal is to create more books and guides that continue to deliver great value to readers like you for little monetary costs. Thank you again for purchasing this study guide and I wish you the best on your future endeavors!

- Dr. Holden Hemsworth

*Your reviews greatly help reach more students. If you found this book helpful, please leave a review on Amazon, nothing helps more than a few kind words.*

# MORE BOOKS BY HOLDEN HEMSWORTH

*DO YOU NEED HELP WITH OTHER CLASSES? CHECK OUT OTHER BOOKS IN THE ACE! SERIES*

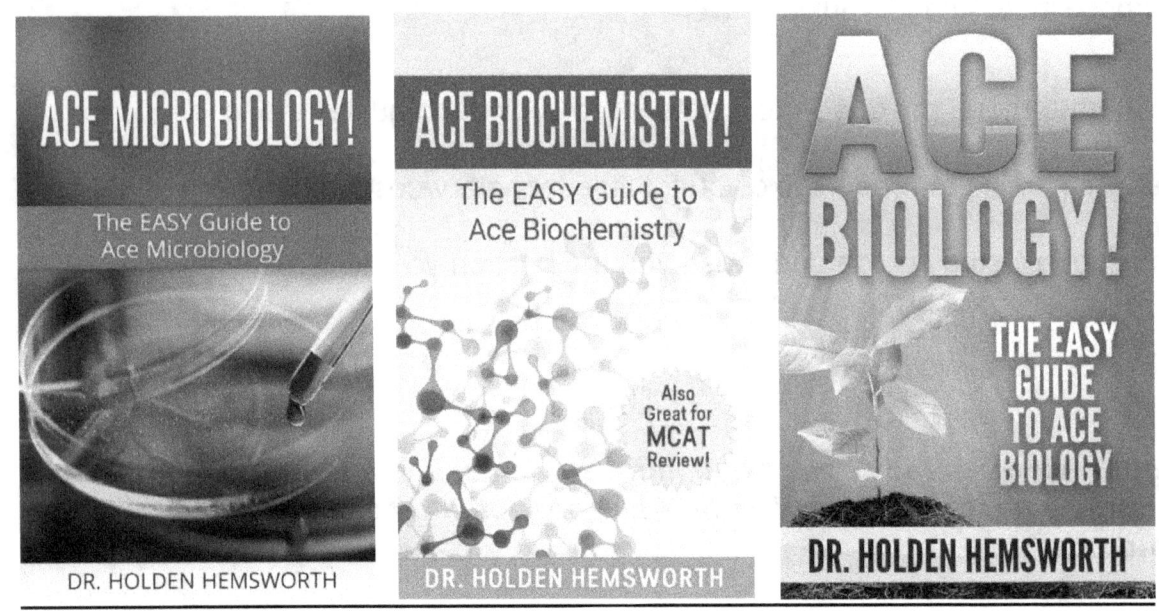

*ALL BOOKS ARE LISTED ON MY AMAZON AUTHOR PAGE*

*MORE BOOKS COMING SOON!*

www.ingramcontent.com/pod-product-compliance
Lightning Source LLC
Chambersburg PA
CBHW080829180526
45168CB00006B/2627